Green Communication
and China

US-CHINA RELATIONS IN THE AGE OF GLOBALIZATION

This series publishes the best, cutting-edge work tackling the opportunities and dilemmas of relations between the United States and China in the age of globalization. Books published in the series encompass both historical studies and contemporary analyses, and include both single-authored monographs and edited collections. Our books are comparative, offering in-depth communication-based analyses of how United States and Chinese officials, scholars, artists, and activists configure each other, portray the relations between the two nations, and depict their shared and competing interests. They are interdisciplinary, featuring scholarship that works in and across communication studies, rhetoric, literary criticism, film studies, cultural studies, international studies, and more. And they are international, situating their analyses at the crossroads of international communication and the nuances, complications, and opportunities of globalization as it has unfolded since World War II.

Green Communication and China

ON CRISIS, CARE, AND GLOBAL FUTURES

Edited by Jingfang Liu and Phaedra C. Pezzullo

MICHIGAN STATE UNIVERSITY PRESS | *East Lansing*

♾ The paper used in this publication meets the minimum requirements of
ANSI/NISO Z39.48-1992 (R 1997) (Permanence of Paper).

Michigan State University Press
East Lansing, Michigan 48823-5245

LIBRARY OF CONGRESS CATALOGING-IN-PUBLICATION DATA
Names: Liu, Jingfang, 1973– editor. | Pezzullo, Phaedra C., editor.
Title: Green communication and China : on crisis, care, and global futures
/ edited by Jingfang Liu and Phaedra C. Pezzullo.
Description: First Edition. | East Lansing : Michigan State University Press, 2020.
| Series: US-China relations in an age of globalization | Includes bibliographical references and index.
Identifiers: LCCN 2019037519 | ISBN 9781611863673 (paperback) | ISBN 9781609176426
| ISBN 9781628954036 | ISBN 9781628964042
Subjects: LCSH: Mass media and the environment—China.
| Communication in the environmental sciences—China.
Classification: LCC GE30.5.C6 .G74 2020 | DDC 363.700951—dc23
LC record available at https://lccn.loc.gov/2019037519

Book design by Charlie Sharp, Sharp Design, East Lansing, Michigan
Cover design by Erin Kirk New
Cover art is "To Green Peacock," by Ella Yiran Liu, age 9
(First place in art competition at Fudan Elementary School, 2018), used with permission.

Visit Michigan State University Press at *www.msupress.org*

ON THE INTERSECTION OF EDGE BALL AND COURTESY:
NOTES ON SCHOLARSHIP IN THE AGE OF GLOBALIZATION

Like America or France or Brazil, China is a nation-state riven with fault-lines along region and race, ethnicity and education, linguistics and libido, gender and more general divisions. Media in the United States tend to portray Chinese society as monolithic—billions of citizens censored into silence, its activists and dissidents fearful of retribution. The "reeducation" camps in Xinjiang, the "black prisons" that dot the landscape, and the Great Firewall prove this belief partially true. At the same time, there are more dissidents on the Chinese web than there are living Americans, and rallies, marches, strikes, and protests unfold in China each week. The nation is seething with action, much of it politically radical. What makes this political action so complicated and so difficult to comprehend is that no one knows how the state will respond on any given day. In his magnificent *Age of Ambition*, Evan Osnos notes that "Divining how far any individual [can] go in Chinese creative life [is] akin to carving a line in the sand at low tide in the dark." His tide metaphor is telling, for throughout Chinese history waves of what Deng Xiaoping called "opening and reform" have given way to repression, which can then swing back to what Chairman Mao once called "letting a hundred flowers bloom"—China thus offers a perpetually changing landscape, in which nothing is certain. For this reason, our Chinese colleagues and collaborators are taking great risks by participating in this book series. Authors in the "west" fear their books and articles will fail to find an audience; authors in China live in fear of a midnight knock at the door.

This series therefore strives to practice what Qingwen Dong calls "edge ball": Getting as close as possible to the boundary of what is sayable without crossing the line into being offensive. The image is borrowed from table tennis and depicts a shot that barely touches the line before ricocheting off the table; it counts as a point and is within the rules, yet the trajectory of the ball makes it almost impossible to hit a return shot. In the realm of scholarship and politics, playing "edge ball" means speaking truth to power while not provoking arrest—this is a murky game full of gray zones, allusions, puns, and sly references. What this means for our series is clear: Our authors do not censor themselves, but they do speak respectfully and cordially, showcasing research-based perspectives from their standpoints and their worldviews, thereby putting multiple vantage points into conversation. As our authors practice "edge ball," we hope our readers will savor these books with a similar sense of sophisticated and international generosity.

—Stephen J. Hartnett

Contents

PART THREE. On Futurity

Acknowledgments

Every book involves many people and much labor, and this book is no exception. The pages that follow evolved out of the second biennial Communication, Media, and Governance in the Age of Globalization Conference, cohosted by the National Communication Association (NCA, natcom.org) and the Communication University of China (CUC), in Beijing, China, in June 2018. That conference provided many of the collaborators in this volume a chance to meet in person. We also appreciated the warm-hearted, hard-working CUC graduate students we met. Phaedra is indebted for the invitation to provide a keynote speech and organize a track on environmental communication, as well as the welcome given to her request to invite Jingfang to become a partner in organizing and editing. The National Communication Association funded Phaedra's international flight to that conference through an Advancing the Discipline Grant, written and administered by Qingwen Dong. We are grateful to NCA and CUC, as well as the leadership of Patrick Dodge and Stephen Hartnett at the University of Colorado Denver and Zhi Li at CUC, for encouraging this international collaboration and doing the extensive labor required to bring us together.

We also are grateful for the FIST grant at Fudan University in Shanghai, which enabled us (Jingfang and Phaedra) to meet in person the year before, in October

2017, when we had our initial conversation about this book. With support from this grant, we also taught an environmental communication course alongside Alison Anderson, traveling from the United Kingdom, and Lin Hong, visiting from Beijing. Three of the authors also presented this work in 2019 in Vancouver at the International Environmental Communication Association (IECA) Conference on Communication and the Environment (COCE). We are excited about the emerging internationalization of environmental communication.

This book wouldn't have happened without all of the internationally forward-thinking authors who contributed to this collection, who kept to our ambitious timeline and shared interdisciplinary insights in the pages that follow. We hope you are pleased with the final collection, and that this work fosters more conversations about environmental communication in our respective home countries and beyond. Michael Warren Cook was an excellent copyeditor in the initial draft of the book manuscript. We are thankful to Michigan State University Press for supporting this scholarship, particularly editor in chief Catherine Cocks, and to Stephen Hartnett for launching this exciting new book series and for careful editing.

Jingfang would like to express her gratitude to all who have inspired, encouraged, and supported her during this book project. In particular, I would like to thank Phaedra for inviting me to coedit this book. Without her, I would not have embarked on such an important project. With her passion, wisdom, and (mysterious) work efficiency, the book seems to have reached the printer in no time. Thanks to NCA, especially Stephen Hartnett, and CUC, especially Zhi Li, for your vision leading the umbrella project. As China strives to be greener in all aspects at this historical time, it is high time to initiate such a meaningful project to encourage much-needed collaborative scholarship in the field of environmental communication.

Editing the book can be challenging at times, given the possible conflicting ideas and thoughts between US and Chinese perspectives involved. For this, I have consulted Hu Huang, G. Thomas Goodnight, Jin Cao, Taofu Zhang, Ye Lu, and Jianbin Jin; thank you for your advice and for being a second pair of eyes for me. I also want to extend special thanks to the field actors who are involved in the practice of environmental protection in China, be they green-minded government officials, environmental journalists, NGOs, or grassroots environmentalists. It has been your willingness to change, your passion, and your hard work that made up the real stories unfolding in this book. Without you, this book would be void of content and meaning, and China would be losing the future.

I want to thank Fudan University and my School of Journalism for providing the needed freedom, environment, and encouragement for me to pursue this academic project and international collaboration; my excellent colleagues (names too many to mention) for being the source of inspiration, encouragement, and support; and my bright students Menghui Wang, Dongxue Zhong, Zhi Wang, Siqi Zhang, Nadia Aleo, and Chao Liang for your help with fieldwork, research, organizing material, and formatting references.

I also owe great thanks to friends and colleagues who helped and supported me in one way or another, sometimes during challenging times, sometimes upon urgent request: Xiaowei Peng, Dahai Fang, Di Wang, Jianhua Yao, Di Xu, Ye Zhou, Jing Wu, Zhenguo Liang, Deshun Zhang, Jinglian Liu, Yaoliang Sun, Hongyan Wang, Ranshu Xu, and Yijing Wang. Finishing the book is impossible without the foundation built from my family. Special thanks to my mother, Songqin Yang, and my sisters Jinglan Liu and Jingmei Liu, for always being there for me with your continuous support and unconditional love. Last but not least I thank my angel Ella, for distracting me with giggles and homework, as well as painting the green peacock used for our book cover with your creative spirit and hope for a greener world.

Phaedra also would like to thank those who have encouraged her along this journey. First, to Jingfang, without whom I cannot imagine taking on this project: your intelligence and good humor have made this book possible. I also would like to thank the camaraderie of Stephen Hartnett and Lisa Keränen, who have been warm and supportive since I moved to Colorado in 2015, including and beyond sharing their knowledge of Chinese politics and culture. CUC was a generous host, and I particularly am indebted to Professor Zhi Li, without whom my partner and child might still be lost somewhere on campus. One unanticipated opportunity was spending time with NCA executive director Trevor Parry-Giles: your diplomacy from Beijing to Boulder have embodied the best of the field. Further, to the aforementioned and all others who traveled alongside my child, with good nature about being assigned food names and band names, I am indebted to your sense of humor, love of delicious plant-based food, and welcoming sense of community.

At the University of Colorado Boulder, I would like to thank Leah Sprain (who also traveled with my family to Beijing and Chengdu) and Tiara Na'puti for encouraging this work through the Center for Communication and Democratic Engagement (CDE), as well as your solidarity and laughter; Dave Ciplet, Michelle Gabrieloff-Parrish, Magnolia Landa-Posas, and Manuela Sifuentes for working together for a better world through the Just Transition Collaborative (JTC); Max Boykoff, Beth Osnes, and Becca Safran, as seemingly tireless, creative climate-communication teachers, researchers, and advocates with Inside the Greenhouse (ITG); Peter Simonson, my departmental chair, for supporting this research and our ongoing conversations about decolonizing rhetoric and culture; Lisa Flores for your friendship, without which Colorado would feel much less livable; Colleen Berry for travel tips; Monica Carroll and Jewel Gurule for helping with reimbursements and logistics; and the graduate students who helped me at different points earlier in the project with copyediting, research, and transnational diplomacy: Logan Rae Gomez, Joanne Marras Tate, Michael Warren Cook, and Samantha West.

Gratitude also is felt for my parents, Vincent and Carmen Pezzullo, for their ongoing support. Finally, I would like to thank my partner, Ted Striphas, and our child, Niko, for journeying to China with me more than once, reminding me to have fun in between the work, and for loving me no matter where my wanderlust may take us.

Introduction

Phaedra C. Pezzullo

"China Is Crushing the US in Renewable Energy"; "US Recycling Woes Pile Up as China Escalates Ban"; "China's New Panda Park Will Be Three Times Bigger than Yellowstone"; "China Rises in UN Climate Talks, While US Goes AWOL"; "Will China's Growing Appetite for Meat Undermine Its Efforts to Fight Climate Change?"; "'Avengers: Endgame' Is Now China's Biggest-Ever Foreign Film"; "In China, 'Cancer Villages' a Reality of Life"; "As It Looks to Go Green, China Keeps a Tight Lid on Dissent."[1]

These headlines characterize how contemporary US media frames China's environmental news. As reflected in this discourse, China plays a significant comparative role in the United States' imagined community, where people increasingly are concerned about environmental conditions in China, the United States, and beyond.[2] Despite the ethnocentricity of much US media, China is regularly given attention, even if the news is sometimes clouded by xenophobia, Sinophobia, and/or self-doubt about whether or not the United States is being left behind by China's accomplishments. Given these mixed feelings of fear and admiration, China often appears as either an eco-apocalyptic nightmare or a green utopic dream.

On the one hand, US media regularly portray China as troubled by environmental crises. The news can feel overwhelming, focusing on intensive air and water pollution, cancer villages, unsafe food, desertification, deforestation, rural poverty, migrant worker precarity, growing population sizes, mountains of garbage, e-waste health risks, and more. Those headlines sometimes are coupled with stories of the Chinese government policing dissent and a diversity of public opinions, linking China's environmental problems with an authoritarian, top-down, one-party government.[3] China, in these instances, becomes a worst-case scenario; for example, a local headline asked: "Compare the Air: Denver's Pollution Can Be Bad, But Is It Beijing Bad?"[4] In these moments, China is imagined as a worst-case measuring stick, a comparative symbol by which one might gauge ecological catastrophe.

On the other hand, China increasingly is recognized as a global environmental leader, as reflected in its rapid renewable energy development, green architectural achievements, and ongoing world climate leadership. US discourse, in these instances, highlights how nongovernmental organizations are mobilizing around anti-pollution and conservation efforts; tour and film companies are selling a longing for China's sublime landscapes; and the government is setting a bold environmental policy agenda. If the current century belongs to Asia rising again, China is a key continental and global leader in futuristic trends and fantasies. As one headline declares: "China IS the Asian Century."[5] In these stories, China is depicted as an emerging global environmental leader with great momentum and a caring commitment to improve environmental regulations, technologies, and values.

Across US media framing, then, China is portrayed as both the worst environmental culprit and the planet's last best hope. Despite these sometimes polarizing and sensational frames, China's environment, as with any nation, lies somewhere in between the extremes of an incessant nightmare or a perpetual dream. Every nation faces challenges related to pollution, environmental justice, climate change, and more—and also has positive stories of hope to share. In the pages that follow, we will move beyond the headlines.

In this volume, we hope to illustrate how environmental communication in China deserves increased scholarly attention in China, in the United States, and beyond—both for being indicative of the contemporary global landscape and for its exceptionality. To do so, we examine China's green communication, from historical values to contemporary ones, from leisure to labor practices, and from advocacy to global policy to art. With China and the United States as our primary audiences, we explore nation-specific and relational environmental expression. Overall, we believe this book will help us consider what contemporary environmental politics entail for "naming, shaping, orienting, and negotiating . . . our ecological relationships" in China, for China-US relations, and across the globe.[6]

Given its geographic scale, economic impact, and population size alone, China is emerging as a significant site for the production of, and research about, the academic field of environmental communication. To address these elements while responding to a crucial gap in scholarly literature on environmental communication in the Global South, this book showcases some of the leading environmental communication research in, with, and about China by international scholars to illustrate the exciting and timely range of topics and approaches emerging from green communication studies. As is apparent from accounts of tourists, consumers, citizens, nongovernmental organizations, and governments, environmental communication in, with, and about China has profound impacts that exceed national borders. Overall, this book illustrates how environmental communication may be studied in culturally specific contexts (such as the nation of China or a subcultural movement in one city), through cross-cultural analysis (particularly framing China-US relations), and as a way to rethink the agenda of the academic field of environmental communication (which tends to focus, although not exclusively, on assumptions inherited from hegemonic Euro-American cultures).

My introduction proceeds in five sections. First, I recount the US-based birth of environmental communication as an area of study. Second, I elaborate on how the ethics of environmental communication research are grounded both in crisis and care. Third, I consider China's potential role in shaping the future of both environmental communication and environmental practices. Fourth, I revisit the state of green politics and public cultures in relation to this research, thus wading into the complicated question of how environmental politics thrive, or not, in contemporary China. Finally, I offer a chapter map, highlighting this volume's overarching themes of care, crisis, and futurity. Throughout, I foreground a sense of cautious hope about China, its evolving green public cultures, and the fate of environmental communication in and between the US and China, as well as beyond.

Environmental Communication

Imagining ourselves as part of a larger environmental community might best begin with acknowledging that *nature can communicate.* In addition to humans, trees talk to each other using soil fungi to signal warnings, search for kin, and share nutrients before they die.[7] Dolphins develop individual whistles as part of their identities and use sounds and nonverbal signals to play, warn, and seek out kin.[8] Rivers have the right to a legal hearing in Ecuador and New Zealand.[9] To address ecological imbalances, therefore, we may consider how nonhuman elements are communicating and/or miscommunicating in banal and extraordinary ways.

Further, the first principle of environmental communication is that *everything is interconnected.* One of the great myths of neoliberal capitalism is that people can survive as individuals in isolation. What ecology teaches us time and time again is that we are relational: humans depend on clean water, air, and food to live, while these basic life elements require that humans operate within ecological limits to be sustained. As the Scottish immigrant and US environmental movement leader John Muir wrote: "When we try to pick out anything by itself, we find it hitched to everything else in the Universe."[10]

In addition to these fundamental communicative capacities and ecological interdependencies, humans have so radically transformed the relationship between nature and culture that we might call it "natureculture."[11] That is, no species or ecosystem is untouched by humans' chemical revolution or what Canadian journalist and activist Naomi Klein names "extractavist" desires.[12] As such, we can find persistent organic pollutants in the Arctic and in our own bodies, no matter where we live. This blurring of boundaries is, in part, why I tend to refer to "environment" more than evoking a more historic, pristine sense of "nature." Further, I follow the lead of US-based environmental justice advocates in prioritizing "environment" over "nature" to signify not only spaces we go to for playing, praying, and learning, but also where we live and work—as well as including humans as part of, not somehow separate from, the environment.[13] The environment speaks both to our precarity in a system that has limits and to our stubbornly resilient embeddedness as part of the environment, no matter where we go or what we do.

With these foundational values in mind, *environmental communication* may be defined as "the pragmatic and constitutive modes of expression—the naming, shaping, orienting, and negotiating—of our ecological relationships in the world, including those with nonhuman systems, elements, and species."[14] *Pragmatic* expression includes instrumental persuasion, such as "have less children," "buy this reusable water bottle," and "support renewable energy policy." *Constitutive* expression is equally ubiquitous, but sometimes more challenging to analyze, as it involves the meaning and values we attribute to particular elements. For example, when Chinese President Xi articulates "green is gold," or when the US Green New Deal introduced by US Representative Alexandria Ocasio-Cortez resolves "to achieve net-zero greenhouse gas emissions through a fair and just transition for all communities and workers," he and they *constitute* environmental policy as a worthwhile path to further economic stimulation, linking to both as necessary for each other.[15] These discourses counter the long-standing industry frame that environmental regulation should be imagined as a barrier to job stimulation, shaping how publics perceive what "green" has signified, and could signify.[16]

Acknowledging the significance of communication therefore requires us to attend to improving significant instrumental practices and to engaging the messy work of defining, shaping, and transforming culture. Given this backdrop, let us now consider how the field of "environmental communication" was born from a US-based perspective. After that, we will return to China.

The US Birth of Environmental Communication

As environmental awareness in the United States grew in the mid-twentieth century, so did interest from communication scholars. Most academics identify five foundational essays by rhetorical scholars published in the 1980s that helped solidify the field of environmental communication. Thomas B. Farrell and Thomas G. Goodnight reconstructed communication surrounding a nuclear disaster to consider technical appeals and how the public was imagined.[17] J. Robert Cox analyzed a line of argument, "the locus of the irreparable," frequently used by the US environmental movement to mobilize advocacy for endangered species or public lands that might be lost forever if action is not imminent.[18] Christine L. Oravec considered US environmental discourse in the early 1900s, particularly appeals to the sublime and debates between preservationists (people who wanted to set aside wild places for low-impact activities) and conservationists (people who wanted to use nature in sustainable ways for natural resources).[19] Finally, Peterson evaluated rhetoric from the 1930s Dust Bowl crisis, addressing public debates about land use and farming.[20] The 1980s also entailed the rise of risk communication, as well as related science communication specialties.[21] The point of recalling these essays is to consider how diverse landmark studies fertilized the field for a wider range of topics and theoretical heuristics to be grown in the future.

The 1990s generated the first US-authored books in environmental communication. Notable texts focused on a range of communication practices, including but not limited to challenges posed by the popularity of oversimplified green discourse, sharing with researchers community public-health concerns over *maquiladoras* or factories at the US-Mexico border, and NGOs staging visual events for the global media, such as how Greenpeace tactically created attention to unsustainable whaling practices.[22] During this time, three edited volumes in environmental communication further marked the growth and institutionalization of this generative field of research: James G. Cantrill and Christine L. Oravec's *The Symbolic Earth*; Star A. Muir and Leah Veenendall's *Earthtalk*; and Craig Waddell's *Landmark Essays on Rhetoric and the Environment*.[23]

The momentum continued to build into the next decade, including the emergence of a journal now titled *Environmental Communication*.[24] Further, Robert Cox, a US-based communication studies professor and former president of the national Sierra Club (the oldest environmental nongovernmental organization in the United States), published the first comprehensive textbook on environmental communication in 2006, which is now in its fifth edition as this book goes to press. Drawing on timely empirical and academic developments, Cox's book surveys a broad range of communication approaches, including interpersonal and cross-cultural communication; organizational communication analysis of nuclear legacies; risk and health communication; science and climate-change communication; green governance and public participation; green media and applied arts; environmental law and policy; environmental rhetoric, media, and cultural studies; environmental journalism; and environmental advertising and public relations. As indicated in these early and trendsetting studies, environmental communication is inherently interdisciplinary in method, foreshadowing the rich diversity of methods drawn on in the following chapters.

Given that similar research, teaching, and institutional trends were growing globally, the International Environmental Communication Association (IECA) was established in 2011; it now serves as the host of a biennial conference that has been held in the United States, Sweden, the United Kingdom, and Canada.[25] Despite these origins and English-language monolingualism, founding members included scholars from Asia (including China), Australia, North America, South America,

Europe, and Africa.[26] IECA continues to promote scholarship, curriculum, artists, practitioners, and organizations that build connections across diverse fields of environmental communication involving a wide range of topics (from conservation to climate justice) and research approaches (from intercultural communication to cultural studies of movements and media). Further, the National Communication Association (based in the United States) continues to host a thriving Environmental Communication Division, and a new division has emerged in the International Communication Association. These organizational developments indicate how the study of environmental communication has slowly become institutionalized in academic organizations, creating the infrastructural support needed for the discipline to grow and prosper—including in China today.

To date, environmental communication includes the entire gambit of environmental topics from oceans to the atmosphere, from invisible toxic pollution to serene mountains, and from human health to the welfare of pandas. Approaches also are as diverse as the rich discipline of communication studies, from interpersonal communication about how much water one uses to shower in a week to climate-science communication about global carbon trends, from analysis of discourse at meetings to involve a community in local environmental planning to art exhibits in globally recognized museums sponsored by NGOs, from infographics of news trends about cancer clusters to public protests to promote stronger environmental laws, from corporate public relations after deadly accidents to park tour guides sharing geological history. This wide range reflects the diversity of voices and media constituting, negotiating, and/or contesting environmental communication. Still, despite the flourishing of environmental communication globally, there is, to my knowledge, no such book as this one, wherein we offer comparative analysis of environmental communication both in and between the United States and China. *Green Communication and China*, therefore, aspires to contribute to this evolving discipline while also adding to it a new sense of global interconnectedness.

The Ethics of Crisis and Care

Although the field of environmental communication is vast, we share two deep ethical commitments within and beyond the boundaries of the United States. Each are worth noting. First, Cox has made a case that environmental communication is a *crisis discipline*.[27] Following mission statements guiding cancer and conservation biologists, Cox argues that we have a duty not only to study ecological crises but also to intervene:

> Implicit in the premises of our scholarship itself, is a set of values that orient our work, require of us its dissemination and implementation, and whose ethical orientation serves as the basis of our recommendations for reform—or abandonment—of harmful or dysfunctional practices. And, it is this "pay-off," or ability of environmental communication scholars and practitioners to contribute to the enhancement of society's communicative competence, that invites an ongoing dialogue about the purpose of the work in which we engage.[28]

Just as cancer scientists are not expected to produce "balanced" research on why cancer might be good, environmental communication scholars need not bother with the unethical impossibility of imagining why ecological degradation might be beneficial. We, therefore, are conducting research not merely to observe the impacts of ecological collapse on our bodies and our networked cultures, but also to make informed recommendations about what we should do in response.

For example, one of the most robust areas of environmental communication research related to China has focused on e-waste—perhaps, in part, because the research speaks to US-China relations in a global marketplace, as well as to persistent power inequities. As US-based performance-studies scholar Soyini Madison has emphasized, the crisis of e-waste raises questions of not only global biopolitics, but also our capacity to discern lies and to imagine more just futures.[29] Finding hope, US-based anthropologist Ralph Litzinger also has written about the grassroots and global coalitions of resistance across environmental, labor, and health NGOs that have aimed, for example, to improve Apple's practices in China.[30] As Madison, Litzinger, and others have made clear, one of the key sites of environmental communication is the intersection of consumerism, waste, technology, and nationalism. While *Green Communication and China* does not tackle this issue, I hope it will offer fertile ground for future environmental scholarship to expand that research, teaching, and advocacy trajectory.

Embracing Cox's ethical standard, I also have maintained that environmental communication should be imagined as a *care discipline*. Just as crises galvanize us, we are motivated by what we love. I argue:

As a "care discipline," environmental communication underscores and values research devoted to unearthing human and nonhuman interconnections, interdependence, biodiversity, and system limits. This means we not only have a duty to *prevent* harm, but also a duty to *honor* the people, places, and nonhuman species with which we share our world.[31]

Most of us, for example, prefer to take vacations in beautiful environments, such as scenic ocean beaches, placid lakes, or sublime mountains. Many join conservation organizations because we love a particular place or species. Some of us care about what is happening at our children's schools because we care about our children and their schools. Caring for the environment is not new. The ancient Chinese concept of *jingwei ziran* (敬畏自然), for example, signifies a reverence for nature. Buddhist beliefs in karma and the value of all sentient life also speak to the value of ecological care. Likewise, indigenous and feminist movements have been advocating for an ethics of care for generations.[32] As these examples demonstrate, different notions of care have long been central elements of philosophies and movements committed to social justice.

Together, then, Cox and I argue that by coupling crisis and care as a dynamic and intertwined dialectic, we can recognize existing and emergent environmental communication across the wide range of affective, physical, and political responses that warrant our attention.[33] While we want to foster dialogue and to encourage research about how to address the challenges presented by climate change and endangered species, we also want to explore cross-cultural beliefs about the deep connections we cherish across human-nature relations. In a sense, one might say that we believe environmental communication should address both the *yin* and *yang* of the universe, the energies of destruction and of life.[34] Indeed, following the insights of Taoist dialectic monism, yin and yang are not imagined in isolation; instead, this philosophical concept, epitomized in the *taichi* symbol, is created through the dynamic relationship between the two. In the case of environmental crisis and care, for example, we easily can recall how a crisis (such as a flood, drought, endangered species status, climate change, and more) can motivate great acts of kindness, a rethinking of a society's values, and a cherishing of what has been or could be lost without environmental intervention. We also all know instances in which encouraging tourists to love areas of natural beauty can lead to overloving an area into a more precarious existence.[35] Many of us also are, or know, parents who became motivated to advocate for solutions to address air or water pollution once their beloved child developed asthma or other health complications caused or exacerbated by pollution.[36] Crisis and care, therefore, circle each other in a precarious dance, as complementary, interconnected, and often interdependent partners.

Given these fundamental touchstones, I now turn to the complex and diverse context of environmental communication in, with, and about China. In particular, I am interested in how certain environmental practices in China might goad global citizens to become more reflective about the taken-for-granted commitment to futurity.

The Future and Environmental Communication in, with, and about China

In the pages that follow, I ask readers to pay close attention to how contemporary Chinese activists, practitioners, and scholars are shaping environmental communication—and vice versa. How do people in China speak for and about nature? Which Chinese values are articulated to ecological goals, when, and why? What are the limitations and possibilities of mobilizing publics for environmental action in China through new media, corporate campaigns, and government initiatives? Are there differences between national and local environmental communication beliefs and practices? As these questions indicate, I begin from the understanding that political life in contemporary China is complicated, varied, full of contradictions and conflicts, and also saturated with hope—this is why the chapters included herein offer such a kaleidoscopic range of perspectives, methods, subjects, and expectations.

What unites this work, in part, is the question of futurity—both for the field of environmental communication and from a planetary perspective.[37] While environmental communication research does not exist tabula rasa (or without a history) and is deeply committed to the ethics of intervening in the present, we also have long been future-oriented. US Indigenous environmental philosophy, for example, often invokes seven generations as a standard of judgment. Sustainability echoes this forward orientation, considering how choices in the present make possible or foreclose our capacity to endure. These global discourses also resonate with China's national discourse, which sometimes emphasizes our collective prospects by intertwining future wealth and accomplishment with environmental policy in ways we might consider as overlapping with and extending US-based narratives of futurity. In Zhiqiang Zhou's landmark review of cultural-studies scholarship from and in mainland China, for example, he calls for fostering political imagination through allegorical criticism, which "is committed to rescuing the liberating significance hidden in everyday life, so as to inspire wishes for [the] future."[38]

One exemplar of the role of futurity in Chinese green public culture might be found in city planning. Environmental-justice and American-studies scholar Julie Sze, who studies global cities such as Shanghai and New York, argues that we should pay more attention to what she calls "eco-desire," which centers the significance of stories and imagination, dreams and desire. Sze suggests that "eco-desire" captures "the fusion of desire, projection, profit, and fun in certain top-down versions of eco-development."[39] This discourse, Sze claims, reflects the ways we can solve ecological crises through pursuing technical advancement and economic growth, while deflecting attention away from questions of governance and unequal ecological burdens.[40] We will revisit Sze's arguments in Andrew Gilmore's chapter, where he explores an imaginary future envisioned for Hong Kong.

As Sze and Gilmore indicate, our thinking about possible environmental futures entails communication that takes our imagination seriously. Drawing from Dilip Gaonkar's understanding of the work of the imagination in shaping public culture, Stephen Hartnett, Lisa Keränen, and Donovan Conley underscore how questions of imagination involve communication:

> Such questions about our shared political imaginings are fundamentally questions about *communication*: about how the United States and China envision each other and how such interlinked imaginaries create both opportunities and obstacles for greater understanding.[41]

Further, Hartnett et al. share Sze's and Gilmore's cautious tone in response to what they call the "technophile dream of happy globalization,"[42] hence cautioning us that our environmental crises will not be solved simply by designing new technologies or opening new markets. As many of the chapters herein indicate as well, environmental communication draws upon studies of dreams, as well as their potential shortcomings. I want to underscore this point: envisioning alternative futures requires *the work of the imagination*, a willingness to ponder new forms of collective life.[43]

Rather than importing theories as if they need not adapt to local cultures, which too often is a colonizing practice, rethinking environmental communication from multinational and transnational perspectives should be encouraged, both by exchanging *and transforming* our thoughts through challenges, adaptations, and reinterpretations. Heeding Kent Ono's warning that creating dichotomies between the "West" and the "East" can reify the inaccurate, rigid, and harmful binary thinking that we have nothing in common, I want to underscore that Chinese green communication practices and theories warrant finding common ground globally and locally while not erasing important differences.[44] To situate these ideas and to contextualize the subsequent chapters, I turn now to two key, interrelated areas: the state and green publics.

The State and Policy as Green Communication

As the two biggest global economies and two biggest producers of carbon dioxide in the world, what China and the United States decide regarding environmental policy has impacts that exceed national borders, from the water that flows across oceans to the climate that touches everything.[45] The individual and collective choices of these two nations, therefore, warrant global attention.

Protecting the environment is a value often shared by many in China today, even if results may be uneven. Consider, for example, Chinese President Xi Jinping's 2017 address at the 19th Party Congress, when he stated: "The modernisation that we pursue is one characterised by harmonious co-existence between man and nature."[46] Likewise, Chinese journalist and filmmaker Chai Jing notes in the 2015 globally watched film *Under the Dome*, which now is banned in China: "The strongest governments on earth cannot clean up pollution by themselves. They must rely on each ordinary person, like you and me, on our choices, and on our will."[47] Both the Party and everyday people criticize some past and present choices in China in order to advocate for stronger environmental protections in the future.

Unfortunately, the president of the United States (as this book goes to press) does not share this perspective, although many residents and politicians do. No matter the results of the 2020 election, the United States long has been and remains a contentious nation for environmental protections. For the purposes of this book, however, I will focus on how this contestation plays a role in relations with China.

Although relations between the United States and China have been strained historically, ever since the "normalization" of relations in 1979 the two nations have evolved deep economic ties, as this snapshot from 2016 attests:

> The world's two largest economies are now deeply intertwined, with bilateral trade in goods and services totaling US$659.4 billion last year—up from barely US$991 million in 1978. China is now America's largest foreign creditor, holding about US$1.8 trillion of US national debt, while American companies have set up more than 20,000 businesses in China.[48]

Further, the environmental stories of these two countries are entangled. Whether or not China is willing to recycle waste from the United States, for example, has a large impact on both nations. Increased cancer rates in China have been traced to US demand for "fertilizers, heavy metals, mobile phones" and more.[49] Likewise, environmentalists in California have tracked carcinogens wafting from China across the Pacific Ocean—in the age of globalization, the faraway is now near, pollution honors no border, crises respect no flags, meaning we are all intimately interconnected. Imagining different environmental futures thus requires us to imagine ourselves as citizens who face both dilemmas and opportunities, crises and care, that transcend traditional nation-states.

Politically, China-US relations have varied as US national leadership has transitioned, and as US President Donald J. Trump has changed his rhetoric. In 2015, prior to the United Nations Climate Change Conference (COP21) in Paris, which I attended as a delegate for the IECA, former US President Barack Obama and Chinese President Xi Jinping announced their bilateral agreement to prioritize climate action.[50] The feeling in the air was optimistic for the most part: after twenty years of negotiations, the world finally was ready to rise to the challenge of climate change. Many climate advocates in North America have begun to echo a commonly invoked Chinese sentiment, repopularized by Naomi Klein, about crisis: it presents both danger and opportunity for our future.[51]

Yet, in 2016, when Trump ran for US president, he campaigned, in part, on fostering an affective feeling of xenophobia against China, tweeting right before election day: "The concept of global warming was created by and for the Chinese in order to make U.S. manufacturing non-competitive."[52] While his presidency to date remains hostile to public-health protections, public-lands conservation, and environmental regulations in general (exemplified by calling for the United States to become the only nation in the world to not support the Paris Accord), Trump's policies in relation to China have been inconsistent to date. He has gone from the aforementioned distrust to admiration during his November 2017 visit back to distrust,[53] including tariff wars as this book goes to press.[54] Further, this xenophobia isn't limited to one person or party in the United States; former US vice president Joe Biden, for example, stated in 2014 in an attempt to bolster US nationalism in the face of Chinese economic growth, particularly in the areas of science and engineering: "I challenge you: Name me one innovative project, one innovative change, one innovative product that has come out of China."[55] As this quotation from Biden indicates, in contemporary American politics, bashing China has long been a bipartisan affair.[56]

Perhaps in response to such discourse from the United States, President Xi stated: "Some countries have become more inward-looking and less willing to take part in international cooperation, and the spillovers of their policy adjustments are deepening."[57] President Xi further emphasized:

> No country alone can address the many challenges facing mankind. No country can afford to retreat into self-isolation. Only by observing the laws of nature can mankind avoid costly blunders in its exploitation. Any harm we inflict on nature will eventually return to haunt us. This is a reality we have to face.[58]

Shortly after, President Xi signed a trade agreement with the governor of the US state of California to address climate change, underscoring his commitment to working across borders with those willing to tackle the challenges and opportunities of addressing humanity's biggest ecological crises to date.[59] Further, since California is the world's fifth-largest economy, ahead of the United Kingdom, India, Russia, and most of the world's nations, the implications of this agreement matters far beyond headlines.[60]

While US-based environmental advocates, local governments, and NGOs struggle to have a greater impact on environmental regulations, the national administration of China has become increasingly committed to the value of environmentalism. Driven at least in part by rising public concerns over environmental degradation, the domestic and global environmental leadership of China has been growing since at least 1973, when the First Environmental Protection Working Conference was held in China.[61] In 2012, at the 18th National Congress of the Communist Party of China, "ecological civilization" or "eco-civilization" formally became one of five national development goals along with economic, political, cultural, and social civilization plans: "The key tenets of ecological civilization include the need to respect, protect and adapt to nature; a commitment to resource conservation; environmental restoration and protection; recycling; low-carbon use; and sustainable development."[62] Although many in China are not yet living lives that reflect eco-civilization, these values already have manifested in thousands of environmental fines for companies, palpable investment in green planning, increased legal standing for NGOs, and much more. "The development of a so-called low-carbon economy (*ditan jingji*) is welcome news," as US-based anthropologist Bryan Tilt notes, "in a country where hundreds of thousands of people die each year from ailments linked to air pollution from fossil-fuel combustion . . . where pollution-related economic losses [can] cut into the nation's gross domestic product, and where aggregate levels of greenhouse-gas emissions are the highest in the world."[63]

As noted, rather than framing environmental regulations as a barrier to socio-economic development, as dominant transnational fossil fuel industries often do, many in China champion the argument that improving environmental conditions is an ideal way to create incentives to invent new technologies, create new jobs, and become a—if not *the*—leading visionary of the world's next primary energy source: renewables. This value system led to the aforementioned articulation or linking of "green" with "gold" as essential to the China Dream. Domestically, the green is gold ideal can be witnessed in China's global climate leadership, in the rapid expansion of renewable energy production, and perhaps is best epitomized in the green architectural triumph of Shanghai Tower. Long a symbol of the future in films and literature, the Shanghai skyline has been transformed by this project. At 127 stories, it is the second tallest in the world, yet it also symbolizes an environmental feat as the tallest green building in the world. It has been certified as LEED Platinum and claims to include the tallest wind turbines, the tallest sky gardens, a second skin to reduce its carbon footprint, an internal recycling system that collects and reuses rainwater, Asia's biggest-capacity diesel generator, and the nation's most advanced energy control center.[64] This does not stand in isolation; Eco-civilization Demonstration Areas exist today throughout thousands of villages and towns in China.[65] These projects and places provide evidence that environmental aspirations in China are transforming the landscape of the country, as well as the cultural value of what the Party calls "eco-civilization."

China's solar boom also reflects the best of the "green is gold" philosophy, installing "more than 34 gigawatts of solar capacity in 2016—more than double the figure for the US and nearly half of the total added capacity worldwide that year."

China, therefore, employs 2.5 million people in the solar sector alone, compared to 260,000 in the United States (a disappointing number hurt by the current administration's focus on bringing back coal jobs).[66] Since the next millennium will not be fueled by the energy sources of the last millennium, renewable energy investments in China bode well for its growing role as a global economic and environmental leader.

Of course, China is not without what Mohan Dutta describes as "ecologies of risk," involving "narratives of dispossession, displacement, environmental degradation, and inequality."[67] The path to modernization often is entangled with ecological and human costs, as we also have witnessed in the United States.[68] US-based environmental policy scholar Judith Shapiro constructs a compelling timeline of environmental troubles since Mao, arguing that addressing issues such as climate change, environmental injustice, and biodiversity loss will require humility, creativity, and a willingness to challenge unsustainable patterns from the past.[69] In another critical perspective on China's practices versus beliefs, Chinese-born Fan Shigang counters the US myth that China is stealing US jobs by chronicling a recent wave of factory closings in the Pearl River Delta region.[70] In short, China's desire to become a global environmental leader may be troubled by the sheer immensity of the challenges it faces locally.

Some of the chapters in this book will highlight similar concerns about barriers to China's ecological, labor, and human rights protections that are key to address for a more sustainable future and are worthy of further discussion. Addressing the possibilities and limitations of environmental communication—balancing our sense of crisis and care, our skepticism and hope—brings us to another important context for this book: publics. In the following pages, I will define the key characteristics of this term and what internationalizing our perspective on environmental communication might do to goad us to reassess our understanding of what constitutes "public" deliberation in the United States, in China, and beyond.

Green Public Cultures

As President Xi stated, we cannot solve the major environmental challenges we face alone, as individuals. One person cannot clean the air of a major city. One person cannot bring back shark populations.[71] One person cannot build the batteries we need to store renewable energy. Although individual choices and innovation matter, even they have a greater impact when many individuals collaborate or make similar choices. Environmental advocates, therefore, need to foster public spaces for education and debate, to encourage a sense that one is part of a collective rather than just an isolated individual, to publicize environmental concerns, and to help imagine solutions. This is why Cox and I argue that it is useful to imagine environmental communication in relation to public spheres, insofar as "legislating, judging, policing, and protecting access to public goods, public speech, public participation, public spaces, public policy, and other elements . . . indicate the health of a democracy."[72] Yet, Chinese government and culture is another important reminder that a US model of democratic governance is not required for environmental communication to exist, and that perhaps the European concept of "spheres" is limiting.[73] Rethinking environmental communication through China, therefore, we can begin to broaden our appreciation for why the field of environmental communication remains deeply intertwined with a sense of collective life, cultures, and networks.

With the advent of new media in the new millennium, many have argued that a green public has reemerged in China in robust and exciting ways. In a foundational essay, Guobin Yang and Craig Calhoun identified three key characteristics of what they call the "green public sphere": environmental discourse, publics, and media. It is worth elaborating briefly on each, as well as extending their argument by placing it in dialogue with research from environmental communication.

Environmental discourse is considered the first key element of a green public, since publics do not exist without engagement beyond an individual. Since Yang and Calhoun define environmental discourse from *Greenspeak*, a book written by psychologists and linguists, their definition focuses on human language, which works well with a traditional Habermasian interpretation of the public sphere as based on conversation.[74] "Chinese greenspeak," they clarify, "includes recent neologisms in the Chinese language such as sustainable consumption, white pollution, eco-centricism, endangered species, animal rights, global warming, desertification, deforestation, biodiversity, bird-watchers, and more."[75] Discourse, however, as the discipline of communication defines it, involves both linguistic and nonlinguistic dimensions: not just the words we use, but nonverbal gestures, images, sensual experiences, proxemics, and more.[76] While focusing on what people say when debating collective life is important, we also must remain attuned to embodied, emotional, and contextual communication systems of expression. To analyze what an NGO campaign is communicating about green peacocks, as my coeditor Jingfang Liu does in this book, involves not only what people say about green peacocks, but also images that circulate, social media networks that circulate discourse, cultural beliefs about green peacocks, whether or not one lives in a green peacock habitat, and more. Throughout this book, then, the notion of *publics* as it relates to environmental communication indicates not only words and narratives, but images and emotions, lived daily experiences, and overarching national narratives—the immense and messy give-and-take of ideas in action.

Still, these multifaceted green public spheres, according to Yang and Calhoun, also require *publics.* By this, Yang and Calhoun "refer specifically to individual citizens and environmental NGOs directly engaged in the production and consumption of greenspeak," though they acknowledge it can involve a wider range of actors, including corporate greenwashing.[77] Since public sphere studies emerged globally in the 1980s, the idea has been contested and retheorized as multiple, overlapping, unequal, and contested expressions. Interestingly, the notion of a public, or multiple publics, hasn't died, perhaps due to the fundamental chord it strikes in terms of trying to imagine how people negotiate collective life through discourse. Thus, for environmental matters involving local, national, and international public goods, public services, and public policy, many of us have continued to find "publics" worthy of our attention. It is no surprise, then, that "in China studies," as Yang and Calhoun emphasize, "the concept of the public sphere has been similarly controversial."[78] Taiwanese-born sociologist Ya-Wen Lei argues that China's public spheres conceptually have synergy with European and US public sphere theory, as they resonate with Chinese analysis of public opinion and discourse insofar as "It is public utterance, expression and formation that makes opinion powerful."[79] Like many others, she argues that the public sphere in China did not arise merely with new media; it, however, has transformed as a result of digital and social media networks. As my coeditor Liu argues, "while the Internet is a must, it is not enough by itself to achieve significant social change in the Chinese environmental movement."[80] Further, Yang has found that Chinese villagers may or may not protest against pollution if the source is fellow villagers, reinforcing how mobilization for environmental advocacy is context-dependent.[81] As these examples indicate, and as the following chapters demonstrate, the concept of the "public" is contested and conflicted, yet also fertile for consideration.

Others in the United States and China have debated whether or not the differences in terms of cultures and government policy in relation to freedom of expression might require us to rethink whether the best term to use is "sphere" or "culture" or another term entirely. Liu believes "public sphere" is the wrong framework to best appreciate the ways people engage and negotiate public culture in China; she and US-based scholar G. Thomas Goodnight have argued that publics in China vary culturally from European and US publics in approaches to voting, legal rights, and more.[82] Rebecca MacKinnon, a US-born journalist and cofounder of the citizen media network Global Voices, also has argued that the government in China has adapted to new media and developed what she calls "networked authoritarianism" to repress any democratization efforts.[83] Further, Italian-born anthropologist Anna Lora-Wainwright has argued that our theories of agency in public spheres should become more culturally sensitive, particularly insofar as Chinese cultural norms about criticism and dissent vary.[84] Likewise, drawing on examples from art and the Internet, Patrick Shaou-Whea Dodge has argued that we should trouble the master trope of "harmony" in China by examining more closely "the censorship-expression dialectic" when studying the cacophony of dissent styles that complicate the possibility of fully closing or controlling the range of public opinions that exist in China.[85]

Further, Hartnett has argued that we might do well to identify tropes from both nations in relation to democratic theory in order to rethink future possibilities. In particular, drawing on rhetorical analysis of contemporary political debates, he has identified narratives of democracy (1) in decline, (2) as chaos, and (3) as hope. In doing so, he argues that a dialogue between the United States and China might involve the former advocating for democracy's benefits (e.g., freedom of speech and freedom of assemblage) without engaging in normative assumptions, while the latter might address its own ideals of stability and harmony without conflating democratic values with chaos.[86] From this perspective, both sides can learn from the other, even as they seek common ground.

Without resolving those debates, it is worth noting how conceptualizing green communication warrants a more robust engagement with embodied, multimodal expression while underscoring our commitment to the value of collective life over increased individualization, privatization, and militarization. Further, my hope is that this volume might help advance the productive exchanges that already have begun about taking a European concept of "spheres" into Chinese culture: is "sphere" vague enough to encompass competing visions of public life? Or should the specificity of Chinese governance and culture warrant a different metaphor, such as "public culture" or something else?

The third key characteristic of a green public, according to Calhoun and Yang, is *media*. They argue that "different types of media (mass media, the internet, and 'alternative media'), differ in social organization, access, and technological features," meaning "they influence the green sphere differently."[87] Indeed, more research is warranted on the various affordances of different media. Certainly, the development of digital media has caused a flood of research in China and beyond on its significance to public networking, debate, and dissent.

Fundamentally, I believe it also is important to underscore how media infrastructure itself does not exist somehow distinctly from the environment. Canadian communication theorist Harold Innis's work, for example, mapped how the railroad and technological media innovations culturally and geographically followed the same historical paths of beaver-fur trade routes and cod fisheries, emphasizing the intertwined paths of nature and communication technologies.[88] That major global cities like Shanghai, Beijing, New York, and Los Angeles were built in relation to waterways is not coincidence, but a sign of how nonhuman elements of the environment provide the conditions of possibility for even our most advanced urban centers.[89] In other words, media is not tabula rasa or a blank slate, but designed in response to, and intertwined with, the environment.

To extend Calhoun and Yang's argument, I would like to draw attention to a fourth characteristic of green publics that is so fundamental it is often left out: the *environment*. Water, landscapes, weather, toxins, population density, climate, and more shape the possibilities of green public life. There is no green public sphere, culture, or collective life without the environment. Given the importance of his work to China, let us remember Karl Marx's critique of mass industrial capitalism:

> Physically humanity lives only on these products of nature, whether they appear in the form of food, heating, clothes, a dwelling, or whatever it may be . . . Nature is (1) a direct means of life, and (2) the material, the object, and the instrument of our life-activity . . . Humanity lives on nature . . . nature is our *body*, with which we must remain in continuous intercourse if we are not to die. That humanity's physical and spiritual life is linked to nature means simply that nature is linked to itself, for humanity is a part of nature.[90]

Given the dangers of alienating ourselves from the necessity of the environment for our existence, and from acknowledging we ourselves are not separate from but a part of the environment, I argue that the environment should be considered a key feature of all communication scholarship.

We all need to drink water, to breathe the air, and to live off the land on the same planet with the same atmosphere. As inalienable necessities of life, the air, water, land, and atmosphere, therefore, fall under what we consider "public goods," which, in turn, require public engagement to value and to govern. US communication scholar Rob Asen emphasizes the dynamic nature of this process of negotiation:

> Public good constitutes a practice of articulating mutual standing and connection, recognizing that people can solve problems and achieve goals—and struggle for justice—through coordinated action. In a networked public sphere, there is no singular, universal public good, but multiple articulations of a public good.[91]

The relationality emphasized in Asen's argument is important to note. Do we imagine ourselves as individual islands or members of larger communities? With so many public goods competing for our attention, which should take priority? And, with the growing challenges we face, how can we coordinate action to improve our lives? In the pages that follow, I hope multiple understandings of the public good and green public cultures become more apparent and worthy of our attention. Indeed, the discussions offered in this book point toward a crucial debate—one central to both everyday life in and extraordinary relations between the United States and China—about how our sense of collective politics is emerging in new ways.

Chapter Map

The chapters in this collection are organized around three themes: Care, Crisis, and Futurity. After previewing each, it is worth painting some broad brushstrokes across the chapters to reflect on what this rich range of contributions indicates about environmental communication in the United States, in China, and beyond.

The first section of the book attempts to bring into focus themes of *care*: animals, places, and activities that bring us joy and happiness. The first chapter is Xinghua Li's cross-cultural, historically grounded study of classic environmental communication concepts: the sublime and *quyeba* ("go be wild"). Considering how nature is collectively imagined in relation to humans, Li unearths cultural values in China and the United States that gave birth to valuing the wilderness and spending time outdoors. Then, she analyzes a contemporary US multinational outdoor retail ad campaign targeted at China's budding outdoor recreation trend. Her chapter provides a landmark analysis of how China and US green cultures may overlap, differ, and shape each other.

Next, Janice Hua Xu studies tourist reactions to the Emei Mountain Monkey Reserve through online TripAdvisor reviews of this popular tourist destination. Xu finds that while Chinese-language reviewers tended to reflect an anthropocentric (human-centered) view instead of the ecocentric (ecologically centered) view of English-language reviewers, they also expressed a sense of awe and admiration for the monkeys. Innovative cross-cultural studies such as Xu's help reveal environmental communication assumptions and norms about human-nature interactions within and across different ways of life.

Finally, Jingfang Liu and Jian Lu provide a compelling analysis of environmental nongovernmental organizations (or ENGOs) in China, outlining how ethics of care and crisis often are intertwined. Liu and Lu extend the emerging and robust area of new media research by focusing on three exemplars: WeChat conversations mobilizing support for the endangered green peacock; Weibo solidarity with an activist who appeared under threat (Lei Ping); and pollution impacts on public health publicized via an online app. Liu and Lu find cultural, emotional, strategic, and scientific appeals performed in ways that might exceed US expectations of environmental communication.

The second section of the book shares a collection of essays that focus on *crisis*. First, Binbin Wang and Qinnan (Sharon) Zhou from the China Center for Climate Change Communication (or "China4C") provide a cutting-edge analysis of data from two previously written reports from the United States and China about public opinion related to climate change. Wang and Zhou provide a compelling series of snapshots about how China and the United States compare and contrast through these surveys from a social-scientific perspective.

The next chapter turns from the global to a local crisis, focusing on how publics network with each other to amplify and strengthen their environmental advocacy. Expanding news-report analysis to include social media platforms and onsite interviews with activists, Elizabeth Brunner considers a controversy surrounding a chemical plant in Dalian, China. Drawing in part on her fieldwork in China, Brunner makes a compelling argument that the field of environmental communication should consider failures in addition to successes in order to gain greater insights into how crises might be addressed in the future. She also introduces the idea of "wild public networks," hence intertwining terms from environmental communication and new social media.

The third chapter on crisis is by Pietari Kääpä, reminding us that environmental-communication values and beliefs are negotiated not just on the ground but also on and around the screen. Drawing on ecocriticism, Kääpä foregrounds the significance of key environmental tropes and policies shaping Sino-US film coproductions. He focuses on two 2018 science-fiction exemplars: *Skyscraper* and *The Meg*. His layered analysis reinforces how the film industry is an important site of research to consider global economic policy and imaginaries of science, technology, and natural resources.

After considering roles of care and crisis in green communication, the third section focuses on futurity. Junyi Lv and G. Thomas Goodnight analyze China's One Belt, One Road (OBOR) initiative to consider President Xi's ambitious global trade route agenda. They examine the implications for Fiji, Tuvalu, and the Marshall Islands, in particular, as islands on the frontlines of climate-change impacts. Hoping to foster a transformative diplomacy, they argue that beginning to imagine climate policies about our future through the perspective of too often marginalized islands has compelling possibilities.

Next, Andrew Gilmore takes readers from the Umbrella Movement of Hong Kong to a blueprint of a place not yet built, an imagined eco-city of hope, known as "Green Mong Kok." Grounded in the striking ecological challenges facing China more broadly and this neighborhood more specifically, Gilmore provides a close, rhetorical reading of a blueprint as public protest and a way to foster public engagement. Overall, Gilmore emphasizes why speculative imaginaries of the future warrant our attention as vital sites of negotiating not only ecological choices but also social justice.

The book ends with a conclusion by my coeditor, Jingfang Liu, and a postscript by Guobin Yang. Both read the threads of commonality among the chapters as tea leaves about the future of environmental communication—primarily in China, but also in the United States and beyond. These chapters aim to set an agenda for classrooms, research, and policy in relation to environmental communication.

Overall, this book displays a wide breadth of the interdisciplinary approaches used in environmental communication, which is not exhaustive but hopes to be generative: journalism and mass-media studies; rhetoric, media studies, and cultural studies; green advertising; intercultural communication; discourse analysis; risk and health communication; science communication; and climate communication. This volume also foregrounds some of the wide range of voices communicating about the environment in various publics: beliefs of everyday people in their daily lives, tourist attitudes, activists and NGO advocacy campaigns, government policy, scientist reports, corporate ad campaigns, filmmakers, and journalists. Green communication also may include voices not foregrounded on the following pages, such as lawyers and judges, SEPA (the State Environmental Protection Administration of China), religious leaders, and animals. In addition to approaches and voices, this book further illustrates some of the vast expanse of topics that may fall under the rubric of "environment," including but not limited to nonhuman animals (peacocks and monkeys); places (parks, squares, and cities); public-health impacts (of smog and GMOs); development (from hydroelectric dam projects to resource wars); technology (nano and energy supplies); and practices (from everyday breathing and outdoor recreation to urban planning and global policies).

There remain endless approaches, voices, and topics to engage related to environmental communication in, with, and about China. Whether motivated by care, crisis, and/or futurity, the pages that follow show that there is much more environmental communication research left to do, to learn from, and to act on. Despite the work left to be done, I hope readers begin *Green Communication and China* with the sense that the future of US-China relations hinges, in no small part, on how well we might bridge our national divides to engage in the shared work of environmental communication.

NOTES

1. Sherisse Pham and Matt Rivers, "China Is Crushing the U.S. in Renewable Energy," *CNN*, July 18, 2017; Adam Allington, "US Recycling Woes Pile Up as China Escalates Ban," *Bloomberg BNA*, February 27, 2018; Jennifer S. Holland, "China's New Panda Park Will Be Three Times Bigger than Yellowstone," *National Geographic*, May 10, 2019; Colum Lynch and Robbie Gramer, "China Rises in U.N. Climate Talks, While U.S. Goes AWOL," *Foreign Policy*, May 7, 2019; Marcello Rossi, "Will China's Growing Appetite for Meat Undermine Its Efforts to Fight Climate Change?," *Smithsonian Magazine*, July 30, 2018 (see also Peter Seligmann, "China Is Filling the U.S.' Leadership Vacuum on the Environment," *New York Observer*, March 14, 2018); Sherisse Pham, "'Avengers: Endgame' Is Now China's Biggest-Ever Foreign Film," *CNN*, May 13, 2019; David McKenzie, "In China, 'Cancer Villages' a Reality of Life," *CNN*, May 29, 2013; Michael Standaert, "As It Looks to Go Green, China Keeps a Tight Lid on Dissent," *YaleEnvironment360* (blog), November 2, 2017, https://e360.yale.edu.

2. On imagined communities, see Benedict Anderson, *Imagined Communities: Reflections*

on the Origin and Spread of Nationalism, rev. ed. (London: Verso, 2006); also see Stephen J. Hartnett, Lisa B. Keränen, and Donovan Conley, "Introduction: A Gathering Storm or a New Chapter?," in *Imagining China: Rhetorics of Nationalism in an Age of Globalization*, ed. Stephen J. Hartnett, Lisa B. Keränen, and Donovan Conley (East Lansing: Michigan State University Press, 2017).

3. For a sample of US media linking Chinese censorship with environmental communication, see Steven Mufson, "This Documentary Went Viral in China. Then It Was Censored. It Won't Be Forgotten," *Washington Post*, March 15, 2015; Sara Hsu, "China Wages War on Pollution While Censoring Activists," *Forbes*, August 4, 2016; David Stanway and Sue-Lin Wong, "Smoke and Mirrors: Beijing Battles to Control Smog Message," *Reuters*, February 15, 2017; Paul Mozur, "Inside China's Dystopian Dreams: A.I., Shame and Lots of Cameras," *New York Times*, July 8, 2018.

4. Jeremy Hubbard, "Compare the Air: Denver's Pollution Can Be Bad, But Is It Beijing Bad?," FOXDenver, May 4, 2018, https://kdvr.com.

5. Zachary Keck, "China IS the Asian Century," *The Diplomat*, July 4, 2013.

6. Environmental communication is defined as "pragmatic and constitutive modes of expression—the naming, shaping, orienting, and negotiating—of our ecological relationships in the world, including those with nonhuman systems, elements, and species," Phaedra C. Pezzullo and Robert Cox, *Environmental Communication and the Public Sphere*, 5th ed. (Newbury Park, CA: Sage, 2018), 361.

7. Peter Wohlleben, *The Hidden Life of Trees: A Visual Celebration of a Magnificent World*, trans. Jane Billinghurst (Vancouver: Greystone Books, 2016).

8. Denis L. Herzing and Christine M. Johnson, eds., *Dolphin Communication and Cognition: Past, Present, and Future* (Cambridge, MA: MIT Press, 2015).

9. See "Legal Arguments for the Standing of Citizens and Nature," in *Environmental Communication and the Public Sphere*, 5th ed., by Phaedra C. Pezzullo and Robert Cox (Thousand Oaks, CA: Sage Publications), 337–357.

10. "John Muir Misquoted," The Sierra Club: The John Muir Exhibit, https://vault.sierraclub.org.

11. Donna Haraway, *Simians, Cyborgs, and Women: The Reinvention of Nature* (New York: Routledge, 1991).

12. Naomi Klein, *This Changes Everything: Capitalism vs. the Climate* (New York: Simon & Schuster, 2015).

13. "For us, the issues of the environment do not stand alone by themselves. They are not narrowly defined. Our vision of the environment is woven into an overall framework of social, racial and economic justice. The environment, for us, is where we live, where we

work, and where we play." Dana Alston, *We Speak for Ourselves: Social Justice, Race, and Environment* (Washington, DC: Panos Institute, 1990), 103.

14. Pezzullo and Cox, *Environmental Communication and the Public Sphere*, 5th ed., 13.

15. *Articulation* here is used in the quotidian sense of uttering or speaking, but also theoretically as the linking of two elements. For more on "green is gold," see the United Nations Environment Programme, "Green Is Gold: The Strategy and Actions of China's Ecological Civilization" (2016), 4, https://reliefweb.int/report/china/green-gold-strategy-and-actions-chinas-ecological-civilization. For more on the "Green New Deal," see H.Res.109—*Recognizing the duty of the Federal Government to create a Green New Deal*, 116th Congress (2019–2020), https://www.congress.gov.

16. This discourse is ubiquitous; for one account of the prevalence in the US, see Jeremy Brecher, "'Jobs vs. the Environment': How to Counter This Divisive Big Lie," *The Nation*, April 22, 2014.

17. Thomas B. Farrell and G. Thomas Goodnight, "Accidental Rhetoric: The Root Metaphors of Three Mile Island," *Communication Monographs* 48, no. 4 (1981): 271–300.

18. J. Robert Cox, "The Die Is Cast: Topical and Ontological Dimensions of the Locus of the Irreparable," *Quarterly Journal of Speech* 68, no. 3 (1982): 227–239.

19. Christine L. Oravec, "John Muir, Yosemite, and the Sublime Response," *Quarterly Journal of Speech* 67, no. 3 (1981): 245–258; Christine L. Oravec, "Conservationism vs. Preservationism: The 'Public Interest' in the Hetch Hetchy Controversy," *Quarterly Journal of Speech* 70, no. 4 (1984): 444–458.

20. Tarla Rai Peterson, "The Will to Conservation: A Burkean Analysis of Dust Bowl Rhetoric and American Farming Motives," *Southern Speech Communication Journal* 52 (Fall 1986): 1–21.

21. Alonzo Plough and Sheldon Krimsky, "The Emergence of Risk Communication Studies: Social and Political Context," *Science, Technology & Human Values* 12 (1987): 4–10.

22. M. Jimmie Killingsworth and Jacqueline S. Palmer, *Ecospeak: Rhetoric and Environmental Politics in America* (Carbondale: Southern Illinois University Press, 1992); Tarla Rai Peterson, *Sharing the Earth: The Rhetoric of Sustainable Development* (Columbia: University of South Carolina Press, 1997); Kevin Michael DeLuca, *Image Politics: The New Rhetoric of Environmental Activism* (New York: Guilford Press, 1999).

23. James G. Cantrill and Christine L. Oravec, eds., *The Symbolic Earth: Discourse and Our Creation of the Environment* (Lexington: University of Kentucky Press, 1996); Star A. Muir and Leah Veenendall, *Earthtalk: Communication and Empowerment for Environmental Action* (New York: Praeger, 1996); Craig Waddell, ed., *Landmark Essays on Rhetoric and the Environment* (Mahwah, NJ: Hermagoras Press, 1998).

24. In 2003, a journal of environmental communication was launched as an annual publication for three years, known as the *Environmental Communication Yearbook*. Subsequently, *Environmental Communication: A Journal of Nature and Culture* was founded and rebranded in 2014 by Taylor & Francis without the subtitle, producing eight issues per year.

25. For more on the IECA, see the website https://theieca.org.

26. "Founding Members," International Environmental Communication Association, https://theieca.org.

27. J. Robert Cox, "Nature's 'Crisis Disciplines': Does Environmental Communication Have an Ethical Duty?," *Environmental Communication* 1, no. 1 (2007): 5–20.

28. Ibid., 6.

29. D. Soyini Madison, "The Mike Daisey Affair: Labor and Performance," *Communication and Critical/Cultural Studies* 9, no. 2 (2012): 234–240.

30. Ralph Litzinger, "The Labor Question in China: Apple and Beyond," *South Atlantic Quarterly* 112, no. 1 (Winter 2013): 172–178.

31. *Oxford Research Encyclopedia of Communication and Critical Studies*, s.v. "Environment," ed. Dana Cloud, vol. 1 (Oxford: Oxford University Press, 2017), DOI: 10.1093/acrefore/9780190228613.013.575.

32. Kyle P. Whyte and Chris Cuomo, "Ethics of Caring in Environmental Ethics: Indigenous and Feminist Philosophies," in *The Oxford Handbook of Environmental Ethics*, ed. S. M. Gardiner and A. Thompson (Oxford: Oxford University Press, 2015): 234–247. See also D. Carbaugh and T. Cerulli, "Cultural Discourses of Dwelling: Investigating Environmental Communication as a Place-Based Practice," *Environmental Communication: A Journal of Nature and Culture* 7, no. 1 (2013): 4–23.

33. Pezzullo and Cox, *Environmental Communication and the Public Sphere*, 5th ed., 16–17.

34. For more, see George Kennedy, *Comparative Rhetoric: An Historical and Cross-Cultural Introduction* (New York: Oxford University Press, 1998); Shaun Treat and Jon Croghan, "Rhetorical Imperialism and Sun-Tzu's Art of War: A Response to Combs' Rhetoric of Parsimony," *Quarterly Journal of Speech* 8, no. 4 (2000): 429–435.

35. See, for example, Roderick Nash, *Wilderness and the American Mind*, 5th ed. (New Haven, CT: Yale University Press, 2014).

36. In the US, for example, advocacy groups such as Moms Clean Air Force and Healthy Child Healthy World. Likewise, in China, journalist and documentary filmmaker Chai Jing recounts her child's health as what motivated her to create *Under the Dome*.

37. On futurity, see José Esteban Muñoz, *Cruising Utopia: The Then and There of Queer Futurity* (New York: New York University Press, 2009).

38. Zhiqiang Zhou, "Problematization and De-problematization—30 Years of Cultural Studies and Cultural Criticism in Mainland China," *Cultural Studies* 31, no. 6 (2017): 782, DOI: 10.1080/09502386.2017.1374427.

39. Julie Sze, *Fantasy Islands: Chinese Dreams and Ecological Fears in an Age of Climate Crisis* (Oakland: University of California Press, 2015), 83, 10.

40. Ibid., 25. Notably, other scholars have been critical of rural ecological construction projects; see, for example, Emily T. Yeh, "Greening Western China: A Critical View," *Geoforum* 40, no. 5 (September 2009): 884–894.

41. Hartnett, Keränen, and Conley, "Introduction," in *Imagining China*, xii.

42. Ibid., xvii.

43. For more on imagination, see Robert Asen, "Imagining in the Public Sphere," *Philosophy & Rhetoric* 35, no. 4 (2002): 345–367.

44. Kent A. Ono, "Preface to Part Three," in Hartnett, Keränen, and Conley, eds., *Imagining China*, 303.

45. According to the Global Carbon Project of the UNFCCC, in 2016, China is the top carbon-dioxide emitter at 10.15 billion tons and the US is second at almost half: 5.31. India follows at 2.3. Chris Mooney and Brady Dennis, "The World Has Just over a Decade to Get Climate Change under Control, U.N. Scientists Say," *Washington Post*, October 7, 2018.

46. Chong Koh Ping, "19th Party Congress: Xi Jinping Affirms China's Commitment on Green Development," *(Singapore) Straits Times*, October 18, 2017, https://www.straitstimes.com. Many quotes could be chosen, like this one: "Lucid water and lush mountains are invaluable assets"; Xing Yi, "Villages Reap Benefits of Ecological Planning," *China Daily*, September 3, 2016.

47. On the global circulation and deep time impact of this now banned documentary, see Fan Yang, "Under the Dome: 'Chinese' Smog as a Viral Media Event," *Critical Studies in Media Communication* 33, no. 3 (2016): 232–244. Quoted here: Anthony Kuhn, "The Anti-Pollution Documentary That's Taken China by Storm," *NPR*, March 4, 2015, https://www.npr.org.

48. Cary Huang, "How the China-US Relationship Evolved, and Why It Still Matters," *South China Morning Post*, November 7, 2016.

49. Anna Lora-Wainwright, *Resigned Activism: Living with Pollution in Rural China* (Cambridge, MA: MIT Press, 2017), 175. For more on media waste and China, see D. Soyini Madison, "The Mike Daisey Affair: Labor and Performance," *Communication and Critical/Cultural Studies* 9, no. 2 (2012): 234–240; Jussi Parikka, "New Materialism as Media Theory: Medianatures and Dirty Matter," *Communication and Critical/Cultural Studies* 9, no. 1 (2012): 95–100; Yu Shi, "Chinese Immigrant Women Workers: Everyday Forms of

Resistance and 'Coagulate Politics,'" *Communication and Critical/Cultural Studies* 5, no. 4 (2008): 363–382; Jingfang Liu and G. Thomas Goodnight, "China and the United States in a Time of Global Environmental Crisis," *Communication and Critical/Cultural Studies* 5, no. 4 (2008): 416–421; and Kent A. Ono and Joy Yang Jiao, "China in the US Imaginary: Tibet, the Olympics, and the 2008 Earthquake," *Communication and Critical/Cultural Studies* 5, no. 4 (2008): 406–410.

50. Tom Phillips, Fiona Harvey, and Alan Yuhas, "Breakthrough as US and China Agree to Ratify Paris Climate Deal," *The Guardian*, September 3, 2016.

51. *This Changes Everything*, Klein Lewis Productions and Louverture Films (2015: Distributed by Abramorama).

52. Donald J. Trump, @realDonaldTrump, 2:15pm, November 6, 2012.

53. When Trump arrived in Beijing, his sentiment changed. See Tony Munroe, "Trump Heaps Praise on 'Very Special' Xi in China Visit," *Reuters*, November 9, 2017.

54. China-US relations have seen troubled waters before. US President Richard Nixon's 1972 visit to China was a diplomatic turning point, as the United States had not previously acknowledged the nation since 1949, when communist leader Mao Zedong began his leadership of the nation.

55. Charles Riley, "Joe Biden Is Wrong. China Does Innovate," CNN Tech, May 29, 2014, https://money.cnn.com.

56. Michelle Murray Yang, *American Political Discourse on China* (New York: Routledge, 2017).

57. Louise Watt, "China's Xi Jinping Decries Failure to Tackle Climate Change with Veiled Attack on Donald Trump," *Independent*, September 5, 2017.

58. Tess Riley, "Chinese President Xi Jinping's Climate Change Remarks Sure Seem to Be Aimed at Trump," *Huffington Post*, October 18, 2017.

59. Associated Press, "China and California Sign Deal to Work on Climate Change without Trump," *The Guardian*, June 6, 2017.

60. Associated Press, "California Is Now the World's Fifth-Largest Economy, Surpassing United Kingdom," *Los Angeles Times*, May 4, 2018.

61. According to the United Nations Environment Programme, "Green Is Gold: The Strategy and Actions of China's Ecological Civilization" (2016), 4.

   Notably, 1970 was a watershed year for the institutionalization of environmental awareness in the United States, including the first Earth Day, the formation of the national Environmental Protection Agency (EPA), and many landmark legislative protections (including the Occupational Safety and Health Act and the National Environmental Policy Act).

62. Xinhua, "Xi Leads Ecological Civilization," *China Daily*, March 22, 2017, http://www.

chinadaily.com.cn.

63. Bryan Tilt, "Dams, Displacement, and the Moral Economy in Southwest China," in *Ghost Protocol: Development and Displacement in Global China*, ed. Carlos Rojas and Ralph A. Litzinger (Durham, NC: Duke University Press, 2016), 87–88.

64. Field notes from author's visit to tower, October 13, 2017. The Platinum LEED status is noted here: Joseph Crea, "World's Second Largest Building, Shanghai Tower, Achieves LEED Platinum," U.S. Green Building Council, December 14, 2015, https://www.usgbc.org.

65. UNEP, "Green Is Gold," 25.

66. Sherisse Pham and Matt Rivers, "China Is Crushing the U.S. in Renewable Energy," *CNN Tech*, July 18, 2017, http://money.cnn.com.

67. Mohan Dutta, "Preface to Part Two," in Hartnett, Keränen, and Conley, eds., *Imagining China*, 168. Brunner also examines the art of Xu Xiaoyan to consider how we might rethink linear notions of "progress"; Elizabeth Ann Brunner, "Contemporary Environmental Art in China: Portraying Progress, Politics, and Ecosystems," *Environmental Communication*, 12, no. 3 (2018): 402–413.

68. Tong has argued that Chinese newspapers tend to frame economic development in tension with environmental protections, particularly emphasizing the correlation between environmental risks and the socioeconomically disadvantaged. Jingrong Tong, "Environmental Risks in Newspaper Coverage: A Framing Analysis of Investigative Reports on Environmental Problems in 10 Chinese Newspapers," *Environmental Communication* 8, no. 3 (2014): 345–367.

69. Judith Shapiro, *China's Environmental Challenges*, 2nd ed. (Cambridge: Polity, 2015).

70. Fan Shigang, *Striking to Survive: Workers' Resistance to Factory Relocations in China* (Chicago: Haymarket, 2018).

71. Thinking beyond the role of just one person, Jeffreys published a compelling cross-cultural article on the ways that translocal celebrity activism has had less of an impact than authoritarian environmental austerity measures in reducing shark-fin luxury food consumption. Elaine Jeffreys, "Translocal Celebrity Activism: Shark-Protection Campaigns in Mainland China," *Environmental Communication* 10, no. 6 (2016): 763–776.

72. Pezzullo and Cox, *Environmental Communication and the Public Sphere*, 5th ed., 24.

73. Many have debated the usefulness of "spheres" compared to other metaphors. For example: Taylor argues for the utility of the concept of "nested public spheres"; Benhabib for a "plurality of modes of association"; Hauser for a "reticulate structure"; and Brouwer and Asen for "public modalities." Seyla Benhabib, "Toward a Deliberative Model of Democratic Legitimacy," in *Democracy and Difference: Contesting the Boundaries of the Political*, ed. Seyla Benhabib (Princeton, NJ: Princeton University Press, 1996): 67–94;

Daniel C. Brouwer and Robert Asen, eds., *Public Modalities: Rhetoric, Culture, Media, and the Shape of Public Life* (Tuscaloosa: University of Alabama Press, 2010); Gerald A. Hauser, *Vernacular Voices: The Rhetoric of Publics and Public Spheres* (Columbia: University of South Carolina Press, 1999); Charles Taylor, *Philosophical Arguments* (Cambridge, MA: Harvard University Press, 1995).

74. Rom Harré, Jens Brockmeier, and Peter Mühlhäusler, *Greenspeak: A Study of Environmental Discourse* (Thousand Oaks, CA: Sage Publications, 1999).

75. Guobin Yang and Craig Calhoun, "Media, Civil Society, and the Rise of a Green Public Sphere in China," in *China's Embedded Activism: Opportunities and Constraints of a Social Movement*, ed. Peter Ho and Richard L. Edmonds (New York: Routledge, 2008), 214.

76. Foucault defines discourse as both discursive and nondiscursive: Michel Foucault, "The Discourse on Language," *The Archaeology of Knowledge and the Discourse on Language* (New York: Pantheon Books, 1972), 215–237.

77. Ibid., 214.

78. Ibid., 213.

79. Ya-Wen Lei, *The Contentious Public Sphere: Law, Media, and Authoritarian Rule in China*, (Princeton, NJ: Princeton University Press, 2017), 16.

80. Jingfang Liu, "Picturing a Green Virtual Public Space for Social Change: A Study of Internet Activism and Web-Based Environmental Collective Actions in China," *Chinese Journal of Communication* 4, no. 2 (2011): 137–166.

81. Yanhua Deng and Guobin Yang, "Pollution and Protest in China: Environmental Mobilization in Context," *China Quarterly*, no. 214 (2013): 321–336.

82. Jingfang Liu and G. Thomas Goodnight, "China's Green Public Culture: Network Pragmatics and the Environment," *International Journal of Communication* 10 (2016), 5535–5557; Jingfang Liu and G. Thomas Goodnight, "China and the United States in a Time of Global Environmental Crisis," *Communication and Critical/Cultural Studies* 5, no. 4 (2008): 416–421.

83. Rebecca MacKinnon, "Liberation Technology: China's 'Networked Authoritarianism,'" *Journal of Democracy* 22, no. 2 (2011): 32–46.

84. Anna Lora-Wainwright, *Resigned Activism: Living with Pollution in Rural China* (Cambridge, MA: MIT Press, 2017).

85. Patrick Shaou-Whea Dodge, "Imagining Dissent: Contesting the Façade of Harmony through Art and the Internet in China," in Hartnett, Keränen, and Conley, eds., *Imagining China*, 315, 313, 333.

86. Stephen John Hartnett, "Democracy in Decline, as Chaos, and as Hope; or, U.S.-China Relations and Political Style in an Age of Unraveling," *Rhetoric & Public Affairs* 19, no. 4

(Winter 2016): 661.

87. Yang and Calhoun, "Media, Civil Society, and the Rise of a Green Public Sphere in China," 213.

88. Harold Innis, *The Fur Trade in Canada: An Introduction to Canadian Economic History* (Toronto: University of Toronto Press, 1930); Harold Innis, *The Cod Fisheries: The History of an International Economy* (Toronto: Ryerson Press, 1940).

89. Notably, New York City has the largest population in the United States and remains one-third the size of Shanghai and Beijing. Other Chinese cities that are home to more people than live in New York City include Tinajin, Shenzhen, Guangzhou, Wuhan, and Chengdu.

90. Marx used "he" and "man" as universal subjects. Given how his theory of alienation impacts all operating in a political economy, I have taken the liberty in this translation to change those terms to "we" and "humanity." Friedrich Engels and Karl Marx, *The Marx-Engels Reader*, 2nd ed., ed. Robert C. Tucker (New York: W.W. Norton & Co., 1978), 89, translation slightly modified.

91. Robert Asen, "Neoliberalism, the Public Sphere, and a Public Good," *Quarterly Journal of Speech* 103, no. 4 (2017): 4.

PART ONE

On Care

Selling the "Wild" in China

Ancient Values, Consumer Desires, and the *Quyeba* Advertising Campaign

Xinghua Li

"**G**o be wild, change your perspective; go at full speed, open your wings and fly," roars Chinese pop singer Tang Siyu in "Go Be Wild" (*Quyeba*), a hit single released in 2012.[1] This was also the theme song for The North Face's eponymous marketing campaign in China during the same year.[2] Against an American-style hard-rock accompaniment, the young man calls on the audience in a coarse voice to "go be wild." He acknowledges the "depression" and "powerlessness" of contemporary urban youth, as well as prescribing the "wild" as the cure: to "turn heaven and earth upside down" and "search for your own paradise." This message offers a double entendre—both to relinquish inhibition or control and to go into the wilderness to perform outdoor adventure sports. Both senses of "going wild" become synonymous in this advertising slogan.

The North Face's "outdoor anthem" epitomizes China's burgeoning outdoor recreation culture. In recent years, dwellers of megacities increasingly suffer from "nature-deficit disorder" and other "maladies of affluence."[3] In response, a new range of outdoor retail companies have sought to market the desire to depart for wild areas for respite and recreation as a type of self-care to recuperate from the stresses of urban life. Coupled with the government's promotion of "leisure culture," the rise of commercial tourism, and the development of infrastructure in remote regions,

3

a young industry of outdoor recreation and nature tourism has begun to thrive in China.[4] Activities such as hiking, camping, biking, skiing, rock climbing, scuba diving, and kayaking are becoming more fashionable pastimes among the young and the wealthy. Their popularity also is owed to diligent marketing. Like their Western counterparts, the Chinese outdoor retail industry—including tour operators, travel media, and outdoor clothing and equipment companies—primarily sells wilderness adventures as opportunities to join the global middle class in enjoying the sun, air, health, and thrills brought by the Great Outdoors.

Yet resorting to the wild as a means of self-renewal is nothing new in China. The country already has a long-recorded history of individuals traveling to nature to seek respite, rejuvenation, and revelations. Daoist mystics and pharmacologists, Confucian poets and literati, and Buddhist pilgrims—to name a few—are known to have roamed such areas since ancient times. Some departed civilization to eschew the corrupt political world and convoluted human relationships, while others sought spiritual transcendence and corporeal immortality in the presence of wild nature. Their journeys and revelations were documented in mythologies, scriptures, travelogues, poems, and paintings and became a part of China's collective memory of wilderness excursions. This chapter compares China's ancient discourses on wilderness excursions with their American counterparts and contemporary outdoor retail marketing to identify a changing subjectivity between body and land, self and world, risk and pleasure.

Wilderness and *Ye*

"Civilizations created wilderness," writes US environmental historian Roderick Nash. Indeed, "wilderness" is a linguistic and cultural construct: it does not have any fixed referents in reality and could be assigned to different objects or places; thus, "one man's wilderness may be another's roadside picnic ground." The idea of wilderness also is constitutive: by distinguishing controlled (domesticated) from uncontrolled (wild) animals, plants, and places, this symbolic construct evokes certain feelings about, values for, and material actions in the natural environment, in turn altering its referents in reality. The early root of the English term "wilderness" is "will," which means "self-willed, willful, or uncontrollable." The adjective "wild," as its derivative, describes the feelings of "being lost, unruly, disordered, or confused." In old English, "wild-dēor-ness" means "the place of wild beasts." Together, the English term "wilderness" refers to uncultivated and uninhabited places, usually with the presence of wild animals, and may make a person feel "stripped of guidance, lost, and perplexed." As these opening comments indicate, "wilderness" is not so much a set place or space as an ancient, complicated, and changing notion constructed via communication.[5]

In Chinese, the character most closely aligned with "wilderness" is "*ye.*" Different from its Old English counterpart, *ye* does not automatically connote emotional responses of fear, perplexity, or loss of control. In pre-Han (before 206 BC) oracle bone script, the character was written as two trees (*mu*) above the earth (*tu*), referring to the border area that led from the farmland into the uninhabited forest. During the Warring States (476–221 BC) period, *ye* indicated "the territory that is beyond administered boundaries (*jiaowai*)."[6] *The Book of Odes* (dated from the eleventh to the seventh century BC) places *ye* in a hierarchy of classified social and natural spaces: "What is outside of the district city (*yi*) is called 'outskirts' (*jiao*). What is beyond these 'outskirts' is called Wilderness (*ye*). External to the wilderness are the forests (*lin*), and still further out is what is called 'arid land' (*tong*)." Compared to *lin* and *tong, ye* is still "a place which can be negotiated for agriculture and pasture."[7] Its ranking in this geo-classification system indicates its subjection to the control of early imperial states. These centralized powers charted space "along an axis emanating from the power center of the court to the margins of the provinces";[8] *ye* was accordingly placed in the borderland as sites of otherness and adventure—but still within control of the empire.

A search in the *Xinhua Dictionary* shows the following entries for the contemporary meanings of *ye*:

> 1) Outskirts, or a place outside the village (as in *yeying*, camping in the wild; *yedi*, wild land); 2) position for politicians that are not currently in power, in opposition to *chao* (as in *chaoye*, court versus exile, *xiaye*, leaving office, *zaiye*, out of political power); 3) unreasonable, impolite, and rude (as in *saye*, throwing a fit, *cuye*, crude or savage); 4) untamed, conceited, with disproportionate ambitions (as in *langziyexin*, wild ambitions of the wolf-let); 5) uncontrollable or hard to control (as in *yexing*, wildness, *zhe haizi xin dou wanyele*, this child has gone wild from playing); 6) animals or plants not tamed or cultivated by humans (as in *yeshou*, wild beast, *yecai*, wild vegetables, *yecao*, wild grass).[9]

While the first and sixth entries can be considered as neutral, the rest all carry derogatory connotations. These dismissive attitudes toward *ye* can be traced to China's mainstream political ideology and civil religion—Confucianism. Originated during the Spring-Autumn Period (770–477 BC), Confucianism proposes the establishment of social order by reinstituting patterns of ritual behavior to restore harmony and humanism. Its general canons, such as *ren* (humanistic altruism), *li* (rites), *zhong* (loyalty), and *xiao* (filial piety) prioritize morality, rituals, and propriety over an individual's spontaneous desire and being. In Confucius's widely quoted phrase "restraining the self and returning to *li*" (*keji fuli*), the self is *an object to be restrained* and *cultivated*.

These connotations have been transferred onto wild places. In the Confucian tradition, wilderness usually was perceived as "not only *jiaowai*, that is, outside the urban outskirts, but also *liwai*, 'outside' or 'ignorant of' any form of rites or socially proper conduct."[10] As Vermeer states, "the Chinese often destroyed forests because they were viewed as 'hideouts for bandits and rebels' and as places where uncivilized people lived."[11] Meanwhile, Confucianism's utilitarian attitude toward nature also permitted large-scale wilderness-transforming activities throughout China's feudal dynasties, such as logging, farming, and damming.[12]

Interestingly, what took place during the feudal history of China bore great resemblance to early American history. According to Nash, when the early European settlers encountered the wilderness in North America, they saw it both as a "cursed and chaotic wasteland" and an immoral and corrupting force "for men to behave in a savage and bestial manner."[13] A tradition of repugnance fueled centuries of destruction of the American wilderness, along with the taming and slaughtering of Native residents. Between the Confucian tradition and the early American pioneer mentality, a parallel may be found: in both cases, wild man and wild land became synonymous as objects of fear and repulsion, as either targets of conquest or agents of destruction.

Discourses of Wilderness Excursions in Chinese History

Historically, activities that ventured into the territory of *ye* varied in purpose, scale, and characteristics of the experience. Since traveling to and surviving in the wild required rich resources and special skills, human presence on these borderlands was largely restricted to indigenous tribes and state-sponsored military activities. *Yeyou* (wilderness excursions), a non-sustenance activity that resembles today's leisurely outings, did exist, but only on a small scale. Emperors and aristocrats were the first to engage in these activities, routinely hosting "hunting trips or imperially sponsored voyages, with the intention of executing important nation- or community-building rituals."[14] Embroidered with elaborate heaven or ancestral worshipping ceremonies, these trips were usually made in the company of a large entourage of servants and bodyguards, who provided physical comfort, and protected the royalty from risks in the wilderness.

The second type of wilderness excursion, nevertheless, was more of a solitary nature and involved a significant amount of physical discomfort and risk. It was mostly associated with religious practitioners, mystics, and pilgrims—and, in the beginning, primarily the Daoists. Developed during the Warring States Period (476–221 BC), Daoism is a religious and philosophical tradition that advocates living in accordance with the Dao—i.e., the path or principle that governs the universe—and engaging in natural and spontaneous actions. Daoism differs from Confucianism in many registers, such as political philosophy, ecological attitude, and cosmology. It is thus unsurprising that Daoist practitioners favored the land rejected by Confucianism, its ideological rival.

In pre-Han China (before 206 BC), wilderness was generally known to be the dwelling place of Daoist mystics and alchemists—"the like of the 'ten sorcerers' (shiwu) or 'wild' men (yeren) who cultivate special arts and . . . are in charge of the 'one hundred drugs' (baiyao) growing in these remote, perhaps semi-mythological areas." They engaged in shamanistic practices and selected wild plant and animal ingredients to make drugs of "immortality."[15] The belief in the medicinal property of wild places and wild beings—China's ancient form of self-care—has its roots in Daoist cosmology. Daoism sees the universe as an energy field (qi) with "ceaseless flows of activity (yang) and receptivity (yin) . . . manifested in the spontaneous arising and decaying of things."[16] The proper path of humans, therefore, is to "model their lives after this natural spontaneity, making ziran, or naturalness, the core value of their philosophy."[17] In Daode Jing, Laozi promotes the principle of wuwei, translated as "non-(artificial) action," with the goal of "achieving the optimal state of harmonic integration between the various dimensions of life." Paradoxically, this "non-artificial action" does not mean "no action." The early Daoists ventured into wilderness to harvest "nature's creative power."[18] They meditated; made medicine, art, and poetry; and studied astronomy, geography, physiology, and proto-chemistry with the goal of cultivating the vital energy (qi) in their body-mind and, hopefully, becoming immortal themselves.

A prominent manifestation of this naturalist philosophy is the Daoist view of the body.[19] Covered by orifices that open and close, the body is believed to continuously receive and expel energy-matter (*qi*) to and from its environment; this interconnection creates a "correlative" relationship, a fractal-like correspondence between the microcosm of the human body and the macrocosm of the universe. Miller explains:

> In the fully realized or perfected (*zhen*) Daoist, the boundaries between self and world are completely porous; it is as though one is fully transparent to the cosmic location in which one is situated. In psychological terms, there is no assertion of ego over and against what is non-ego.[20]

The notion of the "porous" body negates the ego—an imaginary boundary between self and world—and portrays humans as an integral part of the constantly transforming universe. These Daoist beliefs entail a unique set of medicinal and spiritual practices. Traditional Chinese Medicine, heavily influenced by a corresponding view of cosmology-physiology, prescribes long rosters of medicines made of plants and animal ingredients, as well as elaborate sets of body-cultivating (*yangsheng*) techniques such as *taiji* or meditation. The goal is to align "the energy cycles of the body . . . with the seasons, the stars, and the colours in an elaborate and highly technical exercise of meditative harmonization."[21] These teachings, therefore, reflect what Pezzullo in the introduction to this book hailed as driving principles of environmental communication: interconnection and interdependence.

For Daoists dwelling in the wild, the search for human-nature harmonization went even beyond medicine and meditation and concerned the daily routine of survival. For the *xiulian* practitioners, "living in the mountains [was] a highly ritualized and regulated lifestyle." Survival was not a *competition* or *conquest* of the natural environment, but an art of *adjustment* and *alignment* with nature. Also, this alignment required one to sharpen one's senses and elevate them to their finest, subtlest perceptive abilities. Vision was to be enhanced in the cultivating practice (*xiulian*); its energy, however, was *not directed outward but inward*, in the form of "internal vision" (*neishi*). In the Highest Clarity (*Shangqing*) tradition, the most advanced practitioners were said to "feast on pure light energy," and the act of absorbing the light (*fuguang*) caused the practitioner to become a transparent, luminous being—"so luminous and transparent that one could see his internal organs."[22] Corporeal transparency indicated one's achievement of temporal infinity, or immortality, "possibly only because the Dao itself was considered as inexhaustible."[23] With these lines, American readers might think of the transcendentalists—Thoreau, Emerson, Whitman, and their allies—who also shared this dream of the self evaporating into nature. Both American and Chinese cultures, in this sense, share the spiritual tradition of shedding the body-ego and becoming more open and transparent to nature.

Similarly inhabited by Dao, wilderness also bears the attributes of "infinity." Among various types of wild places, mountains occupy a preeminent status in Daoist writings. Often deemed as sacred spaces, they were imagined to be the home of heavenly gods, immortals, and creature spirits. Kunlun Mountain, the "Home of Ten Thousand Gods," for instance, is the seat of the Queen Mother of the West (*Xi Wangmu*), "the starting point of all manifest physical space in Chinese thought."[24] It is also where she places her gardens of the Peach Trees of Immortality as well as many other sacred plants and animals. These mythical beliefs attracted generations of artists and authors who wandered into these mountains. Tao Hongjing, the famous Daoist pharmacologist, alchemist, calligrapher, and poet during the Southern Dynasties (420–589 AD), wrote about his excursions to the Kunlun: "I roam the Sovereign's mountain. All is deeply perfect. The peaks are jagged and high. They are spirits in themselves."[25]

Mark Elvin describes this as "nature as revelation," a philosophical attitude that views the landscape as manifestation of the deepest forces in the cosmos:

> The eye endowed with understanding could see in a landscape the self-realizing patterns of the Way [*Dao*], the ever-renewed cycles of the complementary impulses driving the world's changes. It could divine the geomantic fields of force in protective mountains and power-concentrating pools of water. It could perceive it as the serious playground, so to speak, of the Immortals who were also at the same time its constituents.[26]

Elvin makes two important points: First, "nature as revelation" is an elite perspective: while the majority of the Chinese imperial society were busy clearing forests, damming rivers, exploiting and adapting the natural environment, the attitude promoting reverence and appreciation for nature was largely held by the aristocratic and literate classes (also true in the case of the United States, as Nash argues). Second, in China, this attitude is not exclusively Daoist, but incorporates other schools of thought, such as Buddhism and Confucianism. While Buddhism often found "the embodiment of the Buddha whose spirit radiated, detectably, through all phenomena,"[27] Confucianism also had a tradition of discovering in nature "scenes of moral symbols that would illuminate the ideal qualities of the Noble Man."[28] Robson rightfully notes: "No one sectarian lineage or religion can claim a monopoly on Chinese sacred mountain space; each tradition added its own layers of significations and interpretations to the steadily accreting wealth of religious meaning."[29] It would be fair to say, therefore, that Daoism, Confucianism, Buddhism, and other folk religions fused over time into a set of nature aesthetics that shaped Chinese green public cultures.

This set of aesthetics crystallized, according to Elvin, by the middle of the fourth century, during the Wei-Jing (220–420 AD), Northern (386–581 AD), and Southern Dynasties (420–589 AD).[30] Sandwiched between the unified Han (206–221 AD) and Sui-Tang (581–907 AD), these three hundred years—also known as the Six Dynasties—amount to the longest period of political fragmentation and instability in Chinese history. At a time of national turmoil, local aristocratic culture rose up and elites searched beyond Confucianism for universal meanings. Inspired by Daoist and Buddhist ways of spiritual cultivation, "mystical learning" (*xuanxue*) and "pure discourse" (*qingtan*) emerged to raise larger metaphysical questions "about time, space, human emotions, and mortality" and to "redefine a cosmology built on the responsive relationship between man and the natural universe."[31]

Meanwhile, local aristocrats turned to wilderness as a place of congregation and dwelling. From Wang Xizhi (303–361 AD) of the Orchid Pavilion Gathering, Tao Yuanming (365–427 AD) of the "Record of Peach Blossom Spring," to Xie Lingyun (385–433 AD) of the "Rhapsody on Dwelling among the Mountains," writers, poets, and musicians frequently gathered in mountains and lakesides; composed essays, poems, songs, calligraphy; and drank wine. Many were ousted politicians disillusioned with court politics. Wilderness, for them, became not just a source of spiritual inspiration but a place of political refuge, a site to project their utopian fantasies. Take, for example, Tao Yuanming, a prestigious poet from the Eastern Jin Dynasty (317–420 AD). He resigned from political office during his middle age and returned to an impoverished life of farming and excursions.[32] In the "Record of Peach Blossom Spring," he created a fictional village tucked away in pristine nature, where residents lived an ideal life of simplicity and harmony; this essay has since become the quintessential example of utopia in the Chinese literary canon.

Tao's writings exemplify an ancient hermit culture that was returning to fashion. A brainchild of this culture was "hidden-escape poetry" (*yinyi shi*): poets usually described the natural scenery, the lore and metaphors associated with it, and marveled at the spiritual and aesthetic elation it brought them. Within this genre, a critical trope is *debasement of the ego.* Tao compared himself to wild creatures: "I amount to less than a high-flying bird, seen when gazing up at the clouds. I am nothing compared to a swimming fish, seen when, by a stream, I glance down."[33] Emphasizing the smallness of the self is reminiscent of the early Daoists, who, as mentioned above, sought to fill the body with energies from the environment and become transparent to nature. The difference, however, may exist in the less-than-honest way the Six Dynasties poet-nobles "erase" the self: in the name of "hiddenness" and "escape," many pursued "conspicuously visible withdrawal from the world [with] even a degree of competition for the most distinguished modesty and anonymity." The conspicuousness makes their ego erasure incomplete and renders some susceptible to the accusation of "aristocratic snobbery."[34]

The aesthetics of hiding in nature was further manifested in the "mountain-water" (*shanshui*) paintings.[35] Emerging during the Six Dynasties and thriving through Sui (581–618 AD), Tang (618–907 AD), Song (960–1279 AD), and even into the late imperial times, these ink paintings depict panoramas of natural landscapes *almost* devoid of human presence. They use varying brushstrokes and ink techniques to portray mountains, lakes, springs, trees, and rocks in ethereal spatial and light configurations and under ambiguous weather conditions. Human beings usually are present but appear diminutive compared to the landscape, often pictured as tiny figures steering a raft down the distant river or insinuated by a thatched hut tucked in misty mountains. The size contrast creates the effect of the sublime, which dwarfs the humans and humbles the viewer.

A similar notion of the sublime is a key term in US-based studies of environmental communication.[36] For example, Christine L. Oravec identifies three primary elements of how American environmental leader John Muir invoked the sublime: "the immediate apprehension of a sublime object; a sense of overwhelming personal insignificance akin to awe; and ultimately a kind of spiritual exaltation."[37] Kevin M. DeLuca and Anne Demo study the notion of the sublime portrayed by nineteenth-century American landscape photographers such as Carleton Watkins, who depicted the American West as a pristine and sacred wilderness devoid of human presence.[38] In the same trope, the Hudson River School painters filled their landscape compositions with "precipitous cliffs, dark gorges, and surging storm clouds," while "omitting any sign of man and his works or reducing the human figures to ant-like proportions."[39] During this defining time, oil painting and photography often were used to create a more perspectival and realistic image of specific views in specific US locations during specific times of the year. Even while deploying a new sense of realism, these works offered visions of sublime scenes wherein humans were minor participants.

In contrast, Chinese visual artists often used an "ink and wash" method, which rendered the image monochromatic, two-dimensional, and sparse; their paintings were not meant to reproduce the exact appearance of nature, but rather to grasp a subjective emotion or atmosphere. Consider *Early Spring* by the Song (960–1279 AD) painter Guo Xi.[40] The ink-sketched mountains stand in contrast to the small and distant temples hidden halfway up the hill and appear enormous. The artist inserts a set of mists or clouds—represented by negative spaces—between the foreground and the background, and uses varying strokes for objects in the front and in the back to create "an illusion of depth."[41] The negative spaces block out parts of the viewer's vision and create a gap to be filled in. They, in effect, amplify the sublime by allowing the viewer to imagine the vastness of the mountains beyond positive representation. This configuration reflects the key principles of the Chinese landscape aesthetic. On one hand, it seeks a reflection of the internal, mental landscape of the painter: "Unless there are hills and valleys in your heart as expansive as immeasurable waves, it will not be easy to depict it." On the other, it aims to spread the mental states of the painter to the viewer—by means of negative representation—to manifest a spiritual pursuit of the Dao. "As for landscape, it is substantial, yet tends toward the ethereal plane . . . Landscapes display the beauty of Tao through their forms and men delight in this."[42]

These contrasting schools of landscape artworks were created during times of great political and economic differences between the United States and China. The outwardly directed American vision was largely concerned with discovery, exploration, and settlement, and—even when celebrating the sublime—was steeped in the ideologies of colonialism, nationalism, and modernity.[43] The Chinese paintings, however, primarily reflected an inner pursuit of some ineffable universal order—in a quest for depth and stillness, but not expansiveness and speediness. In a sense, it could be used to explain China's isolationist policies during its last feudal dynasties and the elites' preoccupation with maintaining domestic stability and imperial reputation rather than outward colonialism, as in the Zheng He expeditions. All in all, these historical perspectives will prepare us for understanding the permutations of *ye* in contemporary popular cultures.

Outdoor Recreation and "Donkey Friends" in Contemporary China

How, then, do the contemporary Chinese imagine their wilderness excursions? To answer, I would like to transport the readers imaginatively to the Shanghai metro in 2018, one of China's busiest metro systems with over 10 million daily passengers. Here, a series of public service ads (PSAs) run by the municipal metro company are displayed conspicuously in front of the hustling and bustling crowds.[44] Each ad features an athlete—mostly Caucasian—engaging in outdoor adventure sports such as skiing, rock climbing, and mountain biking. The captions read:

Mountain / is for conquering / for striding over / stomp danger and obstacles under your feet / I am bravery;

Cliff / is for climbing / upward motion / grab gravity in your hand / I am tenacity;

Road / is for trekking / forward motion / throw hardship behind your back / I am determination.

These PSAs aim to boost public morale by promoting qualities of "bravery," "tenacity," and "determination," as well as a Western active outdoor lifestyle. While this is not a notion of self-care involving respite, it does promise that if one can physically conquer hardships in the wild, one might readily feel bravery and tenacity elsewhere in life. In addition to lifting the human spirit, these ads also portray wilderness as a site for recreation and adventure, an obstacle to be conquered, and a venue to assert oneself. The language of exploring and conquering nature gives a feel of Western (American) frontier mentality, and the references to self-affirmation and self-discovery suggest a streak of modern individualism and Romanticism.

How has this imaginary of the wild ascended to prominence in China's green public culture? To answer this question, one needs to first examine the rise of modern tourism in the People's Republic of China.

In most societies of the world, as Urry and Larsen have observed, premodern travels were mostly the preserve of elites, but mass tourism, which democratizes the freedom to travel, is a modern phenomenon.[45] The development of the PRC's tourism industry had a relatively late start. Under Mao's regime, people's freedom to travel was limited, and travel for pleasure was strictly prohibited.[46] It was not until the late 1970s, after Mao's death, that the government began to view tourism as an opportunity for cultivating both economic development and national pride.[47] Initially, China's tourism industry targeted international travelers, but at the end of the 1980s, it turned toward the domestic market.[48] A mass tourist industry took form, centering on packaged tours with a fixed itinerary. Destinations were often constructed into "scenic spots" (*jingdian*)—enclosed and staged places built around a canonic set of "views" (*jing*). As Nyíri notes, this format was a revival of China's premodern literati travel model—a practice started in the Song Dynasty (960–1279 AD) onwards to make rounds of scenic spots cataloged and celebrated in classical writings and religious legends. Post-Mao mass tourism resumed the same format but expanded the scale. Participants traveled in groups, spent long hours on buses, and got delivered to crowded destinations for photo opportunities. Some jokingly describe this experience as "sleeping when on the bus, photographing when off the bus" (*shangche shuijiao, xiache paizhao*).

Nyíri views China's revival of the premodern literati travel as indicating its lack of the "distinctly modern, romantic, exploratory, and self-bettering discourses of tourism" that emerged in the West.[49] Originating in Europe at the end of the eighteenth century, the Romantic ethic shares the same roots with the American transcendentalists; it reacts to the Industrial Revolution with disillusionment and disgust, and expresses the desire to return to nature, tradition, and an individual's inner emotional life. In Western travel cultures, it is expressed as anti-mass-tourism trends that emphasize "the individual rather than the collective experience, exotic locations rather than conventional tourist destinations, small niche suppliers rather than mass . . . operators."[50] Urry and Larsen call this the "romantic gaze," which seeks privacy, solitude, and "a personal, semi-spiritual relationship with the object of the [tourist] gaze." No longer content with *seeing* the sites, the new tourists also look to experience the landscape "bodily, sensuously, and expressively" by walking, climbing, or cycling through it.[51] Adventure travel and sports are part and parcel of this experiential turn.

By comparison, Nyíri argues, Chinese mainstream tourists have yet to develop the desire for privacy, solitude, and intimate interaction with the host environment in comparison with the West. Still trusting in cultural canon and authority endorsements, many do not share the "good" individual travel versus "bad" mass-tourism dichotomy and are generally content with the collective experience of packaged tours. While Nyíri's argument about China's mass-tourism culture may be correct, it does not preclude the emergence of countercultures. In the late 1990s, for example, groups of young people on Chinese Internet travel forums such as Sina.

com started to discuss and organize a new way of travel—backpacking. Already popular in the West, backpacking avoids mass-tourist infrastructures and goes "off the beaten path" to visit underdeveloped, wild areas to hike, trek, or camp.[52] The Chinese Internet community quickly adopted this practice. Members used public forums, and later their own websites, to organize their trips and find like-minded travel companions.

It was on these websites that the "donkey friends" (*lüyou*) phenomenon emerged. Roughly an equivalent to the English term "backpacker," "donkey friends" refers to independent Chinese hikers and adventure tourists who are "largely urban-based, upwardly mobile professional adults."[53] "Donkey" (*lü*) is a homonym for "travel" (*lü*); the image of donkey is a playful simile to describe the backpackers themselves—"carry[ing] provisions and equipment on their backs, plodding along the trail like a donkey."[54] Donkey friends take pride in the pursuit of hardship and danger; they make an effort to distinguish themselves from conventional tourists who mostly seek comfort and pleasure. Obstacles, for them, are means to seek the authenticity of experience and the actualization of the self. Meanwhile, these quests also "co-exist with the pursuit of hedonism and play." Some jokingly classify their enjoyment into two types—"masochism (*zinüe*)," physically challenging activities such as hiking and climbing, and "debauchery (*fubai*)," pleasurable activities such as eating in restaurants, drinking in bars, dancing, etc.[55] The logic is that one must endure "masochism" to earn the right to "debauchery."

While tourism in China has long been viewed "as a modern Western practice," many donkey friends see themselves as treading the paths created by the West, rather than trailblazing themselves.[56] In a Shanghai-based travel magazine, for example, a Chinese woman wrote about her trip to India where she joined the international backpacking community; she described backpacking as an embodiment of "a nation's economic strength" and admitted the "feeling of national pride" when Americans and Japanese admired her for traveling alone.[57] Rather than a solitary subject who *explores* unknown land, she views herself as *retracing* the path of modernity that the West has already opened—but is proud to be a Chinese citizen to do so.

Of course, this path to modernity is not accessible without the "North Face fleece jackets and Timberland boots and backpacks."[58] Many donkey friends place great trust in purchasing and utilizing outdoor clothing and equipment, which are by no means cheap. In an interview with *China Youth Daily* in 2002, a mountaineer talks about the cost of his lifestyle:

Those mountaineering equipment are *all imported from foreign countries* [emphasis added] . . . Climbing a mountain once costs 7–8,000 [RMB]; a whole set of gear usually exceeds 10,000 (RMB); there are even more equipment for caving; you can't even talk about photography—if you are an enthusiast, then it is limitless.[59]

Identifying these goods as imported testifies to the perceived authority of the West in the choice of outdoor gear. By the time the donkey friends were exposed to backpacking in the late 1990s, most international outdoor brands had yet to enter China. Members found English-language adventure shows and outdoor survival manuals on the Internet and circulated them on the public forums. They extracted and translated the technical information, and even created their own tutorials.

A graph from a travel blog on Sina.com shows the elaborate gear set that a beginner hiker "should" acquire.[60] It pictures a human figure wrapped by specialty clothing and gadgets from top to bottom—hat, sunglasses, face masks, gloves, kneepads, four types of clothes (each defending against wind, rain, storm, and sweat), five types of pants, and three types of shoes. There are also camping gear and navigation tools. The body is armed to the teeth, defended from all angles from the environment, without one inch of skin showing. This image traces the *insulated ego* of the outdoor adventurer—one that is protected from any discomfort and danger in the wild, while maximizing pleasure in the quickest, safest, and easiest way. The adventurer's body is no longer a *porous* entity, like that of the Daoist's, but is sealed, water- and dirt-tight, separated from nature.

Since hardship and risk play a critical role in generating the pleasures of an ideal outdoor adventure, the obsession with technology potentially threatens the core of the experience. In the Western context, Beedie and Hudson argue that modern adventure tourism creates "a paradox whereby the more detailed, planned, and logistically smooth an itinerary becomes, the more removed the experience is from the notion of adventure."[61] Unlike traditional adventurers who take years of practice to accumulate the skills to navigate and survive in the wild, the adventure tourist bypasses all of that. The adventure industry also frequently transforms wild places with manmade paths, signposts, viewing platforms, gondola lifts, or even Wi-Fi coverage—all to make the space accessible and inhabitable for humans. This process often has negative impacts on the environment and its plant and animal inhabitants. Quick, safe, easy, and adventure tourism thus "democratizes" the opportunity to access the wild, but it also tames the land and makes obsolete the traditional skills of navigating and surviving in wilderness—an art the American pioneer settlers, the ancient Daoists, among many others, excelled in.

Similarly in China, the transition from a backpacking subculture to mass adventure-tourism culture has been dependent on a process of commodification. Since the early 2000s, the donkey-friends community expanded rapidly. The next generation of "adventurers" began to embrace the whole spectrum of outdoor recreation, from "hard" activities such as scuba diving, mountain biking, white water rafting, and kayaking, to "soft" ones like horseback riding, dogsledding, and wildlife viewing.[62] Their footprints spread from the wilderness in China to the rest of the world. The new societal fascination created a lucrative market for businesses in which hiking clubs, adventure tour operators, and outdoor books and magazines mushroomed. Yet the biggest winners appear to be the manufacturers and retailers of specialty clothing and equipment. Indeed, from 2003 to 2016, China's outdoor apparel and equipment sales rose from 100 million to 23.7 billion RMB, a whopping increase of 2,300 percent.[63] International companies also entered the Chinese market to compete and collaborate with domestic companies. From magazine and metro ads to sponsored documentaries and music videos, outdoor businesses and trade associations rallied media to portray outdoor life as stylish and "cool" while projecting promises of leisure, escape, and self-discovery onto the natural environment. In short, China seemed to have entered the "Outdoor Era."

American Outdoor Recreation Marketing in China

In this section, I will return to The North Face's "Go Be Wild" (*Quyeba*) campaign as an example of how an international outdoor retail company uses a mixture of Western and Chinese images and values to market to a Chinese audience. As an exemplar of these contrasting histories and overlapping contemporary trends, this campaign provides a compelling opportunity to study how environmental communication has been shaped by global consumerism and by the relations between the United States and China.

Founded in San Francisco, California, in 1966, The North Face is a US-based company that manufactures outdoor apparel, equipment, and footwear. Its name refers to the north face of mountains in the Northern Hemisphere, "generally the coldest, iciest and most formidable route to climb."[64] Naming itself after an extreme environment indicates the company's dedication to "extreme" sports—those that challenge, and thus emphasize, the limits of nature and the human body. The North Face entered the Chinese market in 2007, and within four years, its business grew sevenfold.[65] Determined to make China its second-largest market in the world, the company launched a multimedia marketing campaign in 2012 titled "Go Be Wild" (*Quyeba*). The campaign consisted of full-wall advertisements in the busiest metro stations and shopping malls in Beijing and Shanghai, a website and a mobile app, and promotional events cosponsored with outdoor clubs and media. It also recruited singer Tang Siyu and filmmaker Lu Chuan to create, respectively, a "Go Be Wild" theme song and a similarly themed documentary.

Let us first examine the "Turn Heaven and Earth Upside Down to Experience the Feeling" series to promote the Thunder Down Jacket.[66] It is the brand's winter campaign, calling on the "couch potatoes" to "face the cold wind, walk out of the greenhouse, throw oneself into the arms of Nature and turn Heaven and Earth upside down."[67] The leading print ad shows the back of a human figure, wearing a North Face jacket, standing against the sky and holding some rock formation over his head. While strange at first glance, a second look reveals that it is actually an inverted photograph of a rock climber doing a handstand on a snow-covered mountain. Since the image is turned upside down, it creates the illusion that the rock climber is holding up the mountains. The caption reads: "If you feel nothing, don't blame the winter. Go turn heaven and earth upside down. *Go wild.*"

The visual antics of this image call for a comparison with traditional Chinese *shanshui* paintings. As previously noted, the latter frequently depict natural landscapes in sublime proportions; humans appear as small and insignificant, blending into the environment in harmonious fashion. In this image, the human figure stands out sharply from its background—red and black jacket against the blue sky—marking a *clear separation between the ego and the environment.* The size of the body is enlarged to make the mountain range seem small. The *inversion* technique further breaks the sublime effect: it turns the landscape upside down and makes the human the foundation that supports Nature. In addition, this ad resembles a *mid-action selfie* that lets the athlete look back at his own performance through a recording device. The 20/20 vision of the camera allows him to "see" the natural landscape in high definition, with no subjective "distortions" in the field of vision.

In a sense, the clarity and realism hark back to the nineteenth-century American colonialist gaze, and embody the desire to explore an unknown land and thereby conquer or claim it. Yet, the athlete's upside-down perspective portrays an extra sense of "willfulness": it suggests that the landscape can be viewed from *any* angle he wishes, as long as it has *him* as the center. This *willfulness*, or *wildness* (as the etymological root of "wild" is "will") is embodied in the "inverted sublime" that Bell and Lyall characterize: "Nature . . . has evolved from something to look at, to something to leap into, jet boat through, or turn completely upside down."[68] In this regard, the first image amounts to the perfect symbol of global consumerism: narcissism portrayed as adventure, all recorded for social media distribution.

In *Landscape and Power*, W. J. T. Mitchell notes two major shifts in the study of the landscape: the first, associated with modern Western landscape painting (and nineteenth-century landscape photography), aims to view nature via a "transparent eyeball" without any corporeal bias or sensory distortions; the second, associated with postmodernism, decenters the "pure visuality" of the landscape "in favor of a semiotic and hermeneutic approach that treated landscape as an allegory of psychological or ideological themes."[69] Squarely postmodern, the landscape aesthetics of adventure tourism as exemplified here is characterized by two key features: *kinesthesis* and *identity.*

First, according to Bell and Lyall, modern adventure tourism displaces the slow contemplation of nature with a kinetic experience of the body—usually that of acceleration and rush. Rather than "meandering" through the sublime landscape, tourists "accelerate through an increasingly compressed and hyper-inscribed space" and experience maximum stimulation to their sensory organs.[70] For Cater and Cloke, this experience is attractive because it is an "embodied exploration of the self":

> The mystery of how our bodies will respond to the adventurous environment is undeniably part of the attraction. That is, via its elusive nature, the body configures many different spheres of experience. This acknowledgement . . . challenges the phenomenological preoccupation with consciousness as the locus of intentionality.[71]

But this "embodied enlightenment" differs from that of the ancient Daoist meditative practices. While the Daoist practices are slow, quiet, and introspective, adventure sports are fast, noisy, and outwardly projecting. The process inundates one's external senses and discourages the type of inner reflexivity that comes with stillness and quietude. Coupled with the lack of true hardship (built environment) and danger (protective equipment and professional guides), the experience that remains tends to be "a transition from fear to adrenaline-fueled euphoria [, a] progressing from 'AARRGH' to 'YEEHAA!'"[72]

In this world, however, adventure tourists do not travel home empty-handed. A wide range of media technologies (e.g., smartphone cameras, GoPros, GPSs, blogs, social media) are frequently used to record and share these experiences. They function as stand-ins for the introspection and reflexivity that wilderness excursions might have otherwise provided. In the plethora of selfies, GoPro videos, and social media posts, the natural landscape is turned into a system of codified signs circulating in a commodified universe, and at the heart of this commodified universe lies the individual ego.

Consider another print ad, "Living with Madness," issued in Chinese for The North Face's 50th Anniversary.[73] It pictures a BASE jumper, equipped with a parachute, leaping off a cliff into a misty mountain abyss. The camera freezes his body in midair and makes him seem like an immortal walking on clouds. The title calls this action "madness," which, according to the *inverted* logic of postmodern adventure lingo, translates into "heroism." Although the adventure athlete can "walk" on clouds, he is very much *un*like a Daoist immortal, because his glory derives from fighting the sublime rather than succumbing to it. In fact, he participates in dissipating the "myth" of the Immortals. As Bell and Lyall point out, when the tallest peak in the globe has been stood on, "the uninhabited mountain, abode of the gods, could no longer be constructed."[74] Standing atop, the BASE jumper leaps into the previously mystified abyss and defies gravity like a superhuman. His presence in paradise, however, only lasts a few seconds: it is merely in the frozen shots of the camera that he gets to enjoy "immortality"—i.e., "live" perpetually in his perfect form.

So far, the marketing discourse analyzed above mainly targets fans of adventure and extreme sports and is suffused with modern and postmodern Western wilderness values and imageries. The following analysis will engage with a separate set of ads in the "Go Be Wild" campaign directed at the outdoor hiker (i.e., donkey friends) community. They deliberately incorporate elements from traditional Chinese wilderness discourses. This branding strategy was foreshadowed by The North Face's preliminary market research, which found that "appreciating 'the great outdoors' in China is less about extreme sports and more about connecting with nature in a communal, spiritual way." The marketers thus vowed to frame the outdoors in a "Chinese" way—to encourage consumers to "escape the pressures of hectic city life by enjoying the freedom and renewal that comes with spending time outdoors."[75]

These ads coalesce around The North Face–sponsored documentary series "Go Wild!" (*quye!*), directed by the acclaimed environmental film director Lu Chuan.[76]

It consists of four three-minute films, each telling the story of an outdoor lover—"a determined mountain climber, an adventurous volunteer, a passionate wildlife photographer and successful entrepreneur"—and depicting their journeys "redis-cover[ing] themselves and those around them through an exploration of nature."[77] In 2012, print ads introducing these four online films appeared in the metro stations of Beijing and Shanghai, and in the nationally circulated magazine *New Weekly*. These ads gained much wider exposure than the online films themselves, and thus are the subject of this analysis.

This first ad depicts "the wildlife photographer Xi Zhinong, looking for golden monkeys in the Yunnan-Tibet virgin forest."[78] A large yellow tent with The North Face logo is tucked under a snowy winter thicket. A middle-aged man sits inside with his legs crossed, holding a long-distance photo lens. Light shines from behind him into the camera, creating a sense of tranquil beauty. The caption: "*Become the air and the mountain stream. Blend into the green earth. Go dissolve the distance between humans and nature. Go wild.*"

The second ad is about "the founder of the *New Weekly*, Sun Mian, dancing *guozhuo* with Tibetan people."[79] A middle-aged man, clad in a North Face outfit, takes a spin with a few ethnically dressed women and children. A child's body can be spotted behind him, suggesting that they are in the middle of playing hide-and-seek. The man seems happy and relaxed; the ethnic people's faces, though out of focus, appear cheerful and welcoming. "*Get rid of all the tags (you have in the cities). Run freely in the open wilderness. Release yourself and be the wild child of Nature. Go Wild.*"

The third ad portrays "the philanthropist and outdoor hiker Zhang Xiaoyan crossing the Shangri-La Prairie."[80] A female hiker faces away from the camera and strides down a wide cement road that extends into the open horizon. Rows of barley racks, with prayer flags hanging, line both sides of the road. At the far end, Tibetan village houses can be seen at the foot of magnificent mountain ranges, shrouded by mist. "*To embrace the strangeness, togetherness, and separation. Do not wait for anyone. The unexpected is waiting for you on the road. Go wild.*"

The last ad shows the "national mountaineering team ex-coach, Sun Bing, leading the team to ascend the White Horse Snow Mountain."[81] Three mountaineers are trekking up a snowy mountain on a sunny day. The mountainscape behind them is mostly cropped out of the image. The camera zooms in on them wielding hiking sticks and treading on the paths. "*To experience the extreme. Conquer the last meter before the summit. Conquer yourself. Your heart is the wildest mountain. Go wild.*"

Several common themes run across these four ads. First, all of them, to varying degrees, appropriate traditional Daoist ideas and language. In the first ad, the wildlife photographer is seated semi-meditatively in the woods, like an immortal engaging in cultivating practices (*xiulian*); the poetic slogan, "becoming the air/stream/earth," evokes the notion of the "porous" body that absorbs and expels energy to and from its natural surroundings. The second ad presents a successful entrepreneur shedding his urban identity ("tags") to engage in child's play with the Tibetans. It expresses the desire to leave behind worldly achievements and follow the path of nature (it is also decidedly Romantic for lauding the innocence of children and ethnic minorities). The third ad capitalizes on haphazard traveling, a notion that arguably has been long present in Zhuangzi's philosophy of "free and easy wandering" (*xiaoyaoyou*) and Thoreau's sense of walking. The last ad professes, "Your heart is the wildest mountain," referring to a correspondence between inner and outer "landscapes" that traditional *shanshui* painters aspired to achieve. In one way or another, these ads accentuate the theme of dissolving the boundaries of individual ego—by foregoing identity or forsaking intentionality—in search of proximity and alliance with Nature.

In deconstructing the individual ego and personal will, however, these ads also erect a larger national ego in their place. Their visual tropes, in contrast to the verbal slogans, resemble the "role-modeling" method often used in Communist propaganda that depicts heroic and idealistic figures as behavior models for the masses to emulate.[82] All four protagonists are presented in medium shots engaging in exemplary nature-enjoying activities (photographing, dancing, hiking, or trekking); the human figures occupy the spotlight, whereas the natural landscape is either cropped out of the picture or pushed to the distant background. This layout hints at the goal of this documentary series, which is to construct *model consumer identities* to propagate outdoor tourism among the urban Chinese population. Also, all four ads represent Han protagonists exploring Tibetan lands: the first one camps in the "Yunnan-Tibet virgin forest," the second dances "*guozhuo* with Tibetan people," the third and fourth hike the "Shangri-La Prairie" and "White Horse Snow Mountain." The Tibetan landscape, though in the background, appears sacred and mysterious; the Tibetan people, while out of focus, seem happy and innocent, always singing and dancing. Mythologizing the land and infantilizing the locals turns both into readily consumable objects. They endorse the presence and power of the Han tourists in the minority-inhabited border regions and echo the government's goal to modernize these areas through nature and ethnic tourism.

According to Nyíri, the post-reform Chinese state frequently uses tourism to modernize and develop the economy of so-called "backward" areas inhabited by ethnic minorities. Such discourses often highlight the ethnic exotic by "fusing landscape with 'minority' folklore and making spectacles of 'minority customs.'"[83] Yet, behind these exotic enticements, there lies a larger nationalist project to consolidate state sovereignty in the borderlands. In Tibet, Shepherd observes, the government has made systematic efforts to preserve its traditions, especially in cooperating with UNESCO to protect its religious and cultural sites as examples of "world heritage." The efforts present Tibet as a "de-politicized space of 'culture' and 'tradition,'" which allows the government to claim "an imagined unitary national past, thereby justifying contemporary state boundaries and strengthening national standing."[84] This practice stays consistent with the discursive tradition of modern tourism in the PRC, which has been "framed within the fixed boundaries of the nation-state"—one composed of fifty-six ethnic groups sharing a common Chinese heritage and unified under the socialist state.[85] The construction of Tibet's exotic regional identities, in the same vein, insinuates the diverse *and* unified identity of the nation-state.

The presence of donkey friends in Tibet, as promoted by the "Go Wild" documentary series, falls within the same nationalistic frame. Compared to the mass tourists in Tibet, Shepherd notes, the young Chinese backpackers no longer see it as "primitive, dirty, superstitious, and dangerous" but as "exotic, spiritual, authentic and mystical"—i.e., *wild*, natural, and desirable. However, their relationship with the land and the people "remains firmly situated at the level of spectacle." There is little desire to "transcend their positions as observing subjects and enter a backstage into a more authentic zone of cultural reality."[86] In other words, the donkey friends lack empathy to see the world from the Tibetans' eyes, or to view them as autonomous subjects to decide their own fate and future. Many frame their excursions to Tibet, like the abovementioned Chinese female backpacker, as a patriotic act—to explore the nation's vast territory and testify to its geographic, ethnic, and cultural diversity. In fact, the donkey friends' very participation in adventure tourism is enabled by the state's massive infrastructure projects extending deep into the wilderness. As railways, highways, airports, energy pipelines, and telecommunication networks crisscross the Tibetan plateau, Han majority tourists, migrants, and military forces continuously flow into the Roof of the World and help consolidate state sovereignty over this strategic border region.

Considering these political and economic factors, it appears that the modern

Chinese wilderness consciousness, like the US version in the nineteenth century, is also culturally imperialist. While it bears resemblance to some traditional Daoist and naturalist ideas and aesthetics, it often works to mythologize lands and exoticize cultures in the context of tourism to stimulate the consumer desire for the "wild." This commercial impulse plays perfectly into the hands of the state. Adventure tourism not only necessitates the building of invasive infrastructure into China's vast wilderness, but also legitimizes the government's imposition of developmental schemes onto local lands and people in the name of economic growth and national unity building. It perhaps is unsurprising that The North Face capitalizes on this discourse. By calling on the Han tourists to "go wild," its campaign demonstrates a synergy between multinational corporatism and state capitalism—both with colonial roots—to help tame China's remaining wilderness.

Conclusions

To review, The North Face's "Go Be Wild" (*Quyeba*) campaign adopts a wide range of discursive elements to persuade the Chinese audience to buy their products and engage in outdoor recreation. The ads analyzed targeted fans of adventure and extreme sports, diverging in many ways from premodern Chinese *and* modern Romantic representations of the natural sublime. While the latter depict the wilderness as vast and awe-inducing, outdoor adventure sports ads portray wilderness as a backdrop for human activities. The landscape is shrunk to a smaller scale and rotated in different angles to create an "inverted sublime." Always bright, clear, and friendly, the wild landscape is flattened, with its depths filled up, and becomes fully visible to the technologically enhanced eye. Its visibility removes the aura and invites the tourist to approach, conquer, and experience it hands-on. The implication for green public culture is, therefore, a shift from humility to arrogance, from the premodern and modern natural sublime to the postmodern dominated landscape. Rather than "going wild" to become in touch with the environment and enact an ethics of care, consumers are encouraged to "go wild" to prove how exceptional they are as individuals that thrive in a global capitalist world, while being unencumbered by extreme environmental conditions.

The role of the human body in The North Face "Go Be Wild" (*Quyeba*) campaign also differs from traditional Chinese discourses. The latter often represent the body as tiny, porous, and transparent. In contrast, the adventure athlete's body is viewed as larger than life, powerful, with its ego-contour singled out from its environment. Instead of aligning the body with the force of nature, the quintessential model for The North Face sets the human body in opposition to—and in competition with—nature. In wild places, the athletes are always *defying* something: heights, depths, distances, wind, or waves. Contrary to the Daoist practices that are often still and quiet, adventure sports are fast and noisy. The accelerated (but protected) movements stimulate one's external sensory organs to the max, while discouraging introspection and reflection about one's codependency with the environment. Self-care, then, moves from becoming harmonious with nature, to "going wild" as a commercial self-therapy to numb oneself with sensory stimulation and self-worship and unplug from the political and economic consequences of a rapidly consumed Earth.

By comparison, the backpacker ad series of The North Face's "Go Be Wild" campaign more liberally adopts Chinese traditions to interpellate a larger base of consumers. These ads advocate a slower, meditative existence in nature—as in photographing, dancing, and hiking—and encourage one to immerse one's body in wild and unfamiliar environments to practice reflection on both the self and the world. This exemplary behavior coheres with state policies to advance tourism in minority-inhabited regions and the opening up of these regions for recreation of the Han Chinese. The invitation to dissolve the boundaries of the individual ego, therefore, morphs into a demand to adopt a larger national ego. The tourists are encouraged to assume the new, modern standards of national pastime and to visit, experience, and consume the country's most exotic and "wildest" lands and treasures. To sustain this consumerist and nationalist ego, vast networks of highways, railroads, airports, hotels, and cellphone towers are built into China's remaining wilderness; they cut into the flesh of the mountains, rivers, grasslands, fields, and villages and carve them into pieces of commodities for sale on travel websites and in outdoor industry ads.

This playful ethic of postmodern consumerism brings unprecedented challenges to environmentalism. Thanks to postmodern time- and space-compressing technologies (cars, trains, planes, the Internet, outdoor sports equipment, etc.), the new global middle class has turned more and more habitats of others into their "playgrounds." What, then, would be a more ethical way of travel, that is, to journey across the land without crushing the earth? Here, I would like to expand upon Zhuangzi's notion of "free and easy wandering" and Thoreau's notion of "walking." As mentioned, the slogan of the third ad in the "Go Wild" documentary series promotes a form of spontaneous movement based on the unexpected circumstances in the environment, without artificial and goal-oriented planning. This sense of haphazardness in travel, I argue, pays homage to Zhuangzi's notion of "free and easy wandering" (*xiaoyaoyou*). In a number of fables, Zhuangzi used "tropes of floating on the wind or down a river" to convey man's natural, effortless participation in the Dao. To him, a sense of rootlessness is not tragic, but liberating and enlightening, because it accepts and adapts to "the ongoing transformation and bipolar alternation" of the Dao.[87]

Interestingly, Zhuangzi's notion of wandering highly resembles Thoreau's portrayal in his 1862 essay "Walking."[88] According to Thoreau, the most enlightened mode of traveling is walking without technology or infrastructure, and the freest and wildest mode of walking is "sauntering." "Sauntering" derives from the French roots of *sans terre*, "without land or a home," and *à la Sainte Terre*, "to the Holy Land."[89] It implies that true walkers should wander as if one does not have a home or destination; it is then that one can submit to one's internal compass—the call of the "wild"—and be taken to unexpected places with enlightening encounters. Of course, this nostalgic notion of preindustrial and pre-technological wandering could be easily co-opted by the tourist industry into another reason for whimsical yet commodified traveling. But the key difference is, the true form of wandering does not rely on external technology and tourist infrastructure; it heeds the internal voice of the wild, but not the call of Expedia.com or Ctrip.cn. The cross-cultural lessons of wilderness from both China and the United States, therefore, might be that the best type of self-care is practiced not by caring about the self through commodification. Instead, it is achieved by being lost in nature; by embracing its danger, hardship, and otherness; and by risking the ways such adventures might challenge our own limits. Perhaps it is only in losing oneself that one can refind the value of the wild on a higher level.

NOTES

1. Siyu Tang, "Quyeba," on *Quyeba* (album) (Beijing: PianBei Music, 2017).
2. Daniel Gilroy, "The North Face: Inspiring Chinese City Dwellers to 'Go Wild,'" ChinaSmack.com, March 27, 2012, https://advertising.chinasmack.com/2012/the-north-face-inspiring-chinese-city-dwellers-to-go-wild.html.
3. Richard Louv, *Last Child in the Woods: Saving Our Children from Nature Deficit Disorder* (Chapel Hill, NC: Algonquin Books, 2008); "The Maladies of Affluence: Globalization and

Health," *The Economist*, August 11, 2007.

4. On leisure culture, see Jing Wang, "Culture as Leisure and Culture as Capital," *Positions: East Asia Cultures Critique* 9, no. 1 (2001): 69–104.

5. Roderick F. Nash, *Wilderness and the American Mind*, 4th ed. (New Haven, CT: Yale University Press, 2001), 1, 3.

6. Thomas Hahn, "An Introductory Study on Daoist Notions of Wilderness," in *Daoism and Ecology: Ways within a Cosmic Landscape*, ed. N. J. Girardot, James Miller, and Liu Xiaogan (Cambridge, MA: Harvard University Press, 2001), 205.

7. Ibid., 206.

8. Richard E. Strassberg, *Inscribed Landscapes: Travel Writing from Imperial China* (Berkeley, CA: University of California Press, 1994), 10.

9. *Xinhua Dictionary* (Beijing: Commercial Press, 2017), 581–582.

10. Hahn, "An Introductory Study on Daoist Notions of Wilderness," 209–210.

11. Eduard B. Vermeer, "Population and Ecology along the Frontier in Qing China," in *Sediments of Time: Environment and Society in Chinese History*, ed. Mark Elvin and Liu Ts'ui-jung (Cambridge: Cambridge University Press, 1998), 248.

12. Robert B. Marks, *China: Its Environment and History* (London: Rowman & Littlefield Publishers, 2011).

13. Nash, *Wilderness and the American Mind*, 29.

14. Hahn, "An Introductory Study on Daoist Notions of Wilderness," 205.

15. Ibid., 206.

16. James Miller, "Daoism and Nature," in *Nature across Cultures: Views of Nature and the Environment in Non-Western Cultures*, ed. Helaine Selin (New York: Springer, 2003), 394.

17. Liu Xiaogan, "Non-Action (WuWei) and the Environment Today: A Conceptual and Applied Study of Laozi's Philosophy," in *Daoism and Ecology: Ways within a Cosmic Landscape*, ed. N. J. Girardot, James Miller, and Liu Xiaogan (Cambridge, MA: Harvard University Press, 2001), 315–340, quoted in Miller, "Daoism and Nature," 394.

18. Ibid., 394.

19. See figure 1 on GreenAdsInChina.wordpress.com. Due to copyright concerns, all images analyzed in this chapter are published on the author's private blog (http://GreenAdsInChina.wordpress.com). Please visit this blog to access them. The author strongly recommends the reader review these images, which will provide more vivid understanding of the analysis.

20. Miller, "Daoism and Nature," 397.

21. Ibid., 398.

22. Ibid., 212, 401.

23. Hahn, "An Introductory Study on Daoist Notions of Wilderness," 214.

24. Ibid., 206.

25. Ibid., 205.

26. Mark Elvin, *The Retreat of the Elephants: An Environmental History of China* (New Haven, CT: Yale University Press, 2004), 321.

27. Ibid., 321.

28. Strassberg, *Inscribed Landscapes*, 20.

29. Miller, "Daoism and Nature," 402.

30. Elvin, *The Retreat of the Elephants.*

31. Strassberg, *Inscribed Landscapes: Travel Writing from Imperial China*, 26.

32. Ibid., 28.

33. Ibid., 334.

34. Elvin, *The Retreat of the Elephants*, 334.

35. See figure 2 on GreenAdsInChina.wordpress.com.

36. Phaedra C. Pezzullo and Robert Cox, *Environmental Communication and the Public Sphere*, 5th ed. (Thousand Oaks, CA: Sage Publications, 2018).

37. Christine L. Oravec, "John Muir, Yosemite, and the Sublime Response: A Study in the Rhetoric of Preservationism," *Quarterly Journal of Speech* 67, no. 3 (1982): 248.

38. Kevin M. DeLuca and Anne T. Demo, "Imagining Nature: Watkins, Yosemite, and the Birth of Environmentalism," *Critical Studies in Media Communication* 17, no. 3 (2000): 246.

39. Nash, *Wilderness and the American Mind*, 78, 79.

40. See figure 3 on GreenAdsInChina.wordpress.com.

41. Matthew Turner, "Classical Chinese Landscape Painting and the Aesthetic Appreciation of Nature," *Journal of Aesthetic Education* 43, no. 1 (2009): 115.

42. Ibid, 112.

43. W. J. T. Mitchell, "Imperial Landscape," in *Landscape and Power*, ed. W. J. T. Mitchell (Chicago, IL: University of Chicago Press, 2002), 6–9.

44. See figure 4–6 on GreenAdsInChina.wordpress.com. These public service ads are ST Decaux's (the municipal metro company) default posters to hold an ad space before it is rented out.

45. John Urry and Jonas Larsen, *The Tourist Gaze 3.0* (London: Sage Publications, 2011), 5.

46. Wen Zhang, "China's Domestic Tourism: Impetus, Development and Trends," *Tourism Management* 18, no. 8 (1997): 565–571.

47. Honggen Xiao, "The Discourse of Power: Deng Xiaoping and Tourism Development in China," *Tourism Management* 27, no. 5 (2006): 803–814.

48. Khek Gee Francis Lim, "Donkey Friends in China: The Internet, Civil Society, and the

Emergence of the Chinese Backpacking Community," in *Asia on Tour: Exploring the Rise of Asian Tourism*, ed. Tim Winter, Peggy Teo, and T. C. Chang (Oxford: Routledge, 2009), 291–301.

49. Pál Nyíri, *Scenic Spots: Chinese Tourism, Cultural Authority and the State* (Seattle: University of Washington Press, 2006), 58.

50. Urry and Larsen, *The Tourist Gaze 3.0*, 108.

51. Ibid., 19, 111.

52. Xianrong Luo, Songsham Huang, and Graham Brown, "Backpacking in China: A Netnographic Analysis of Donkey Friends' Travel Behavior," *Journal of China Tourism Research* 11, no. 1 (2014): 67–84.

53. Lim, "Donkey Friends in China," 293.

54. Jocelyn Edmund and Gary Sigley, "Walking the Ancient Tea Horse Road: The Rise of the Outdoors and China's First Long-Distance Branded Hiking Trail," *Journal of Tourism Consumption and Practice* 6, no. 1 (2014): 11.

55. Lim, "Donkey Friends in China," 298.

56. Nyíri, *Scenic Spots*, 3.

57. Ibid., 90.

58. Ibid., 89.

59. Ping Cai, "Quitting Jobs to Start Journey: A Group of Nature-Loving 'Donkey Friends' from Online," *China Youth Daily*, April 24, 2002, http://news.sina.com.cn.

60. See figure 7 on GreenAdsInChina.wordpress.com. Retrieved from http://blog.sina.com.cn.

61. Paul Beedie and Simon Hudson, "Emergence of Mountain-Based Adventure Tourism," *Annals of Tourism Research* 30, no. 3 (2003): 627.

62. See "Adventure Tourism," *Encyclopedia of Emerging Industries* (Farmington Hills, MI: Gale, 2018).

63. "2018 Strategic Analysis of Outdoor Sports Market Development in China" (*2018 nian zhongguo huwai yündong de fazhan shichang zhanlüe fenxi*), China Industry Information Research Net (Zhongguo Chanye Xinxi Yanjiu Wang), http://m.china1baogao.com/fenxi/20180316/9627546.html.

64. Retrieved from https://www.thenorthface.com.

65. Jing Meng, "Outdoor Success," *China Daily European Weekly*, August 5, 2011.

66. See figure 8 on GreenAdsInChina.wordpress.com.

67. "The North Face's Fluffy Down Jacket Product Release," Sina Outdoor (2012), http://sports.sina.com.cn/outdoor/2012-09-12/12362516.shtml.

68. Claudia Bell and John Lyall, "The Accelerated Sublime: Thrill-Seeking Adventure Heroes

in the Commodified Landscape," in *Tourism: Between Place and Performance*, ed. S. Coleman and M. Crang (New York: Berghahn, 2002), 27.

69. Mitchell, "Imperial Landscape," 1.

70. Bell and Lyall, "The Accelerated Sublime," 21.

71. Carl Cater and Paul Cloke, "Bodies in Action: The Performativity of Adventure Tourism," *Anthropology Today* 23, no. 6 (2007): 13–16.

72. Ibid., 13.

73. See figure 9 on GreenAdsInChina.wordpress.com.

74. Bell and Lyall, "The Accelerated Sublime," 18.

75. "Growing an American Brand in China: How an Outdoor Retail Company Found a New Asian Audience through Digital and Brick-and-Mortar Shops," Ideo.com (2012), https://www.ideo.com.

76. Lu Chuan became known initially through his small-budget productions, including *Kekexili: Mountain Patrol* (2004), one of China's earliest environmental films about the struggle between wildlife poachers and vigilante rangers.

77. Gilroy, "The North Face: Inspiring Chinese City Dwellers to 'Go Wild.'"

78. See figure 10 on GreenAdsInChina.wordpress.com.

79. See figure 11 on GreenAdsInChina.wordpress.com.

80. See figure 12 on GreenAdsInChina.wordpress.com.

81. See figure 13 on GreenAdsInChina.wordpress.com.

82. Stefan Landsberger, *Chinese Propaganda Posters: From Revolution to Modernization* (New York: Routledge, 1995).

83. Nyíri, *Scenic Spots*, 49, 16.

84. Robert Shepherd, "Cultural Preservation, Tourism and 'Donkey Travel' on China's Frontier," in *Asia on Tour: Exploring the Rise of Asian Tourism*, ed. Tim Winter, Peggy Teo, and T. C. Chang (Oxford: Routledge, 2009), 255.

85. Timothy S. Oakes, *Tourism and Modernity in China* (New York: Routledge, 1998), 45–48.

86. Shepherd, "Cultural Preservation, Tourism and 'Donkey Travel,'" 260, 263, 262.

87. Strassberg, *Inscribed Landscapes*, 22.

88. Henry David Thoreau, "Walking," in *Excursions: The Writings of Henry David Thoreau* (Boston: Riverside Edition, Houghton-Mifflin, 1893), 251–304. See also J. Robert Cox, "*Loci Communes* and Thoreau's Arguments for Wilderness in 'Walking' (1851)," *Southern Speech Communication Journal* 46 (1980): 1–16.

89. Thoreau, "Walking," 251.

From "Charmed" to "Concerned"

Analyzing Environmental Orientations of Wildlife Tourists through Chinese and English TripAdvisor Reviews

Janice Hua Xu

n recent years, China has seen hefty investment in tourism infrastructure to facilitate adventure and nature-based ecotours. In China, tourism has become a strategic pillar industry for the national economy, contributing RMB 8.19 trillion (or 11.01 percent) to China's GDP in 2016. Around 79.62 million people were directly or indirectly employed in the tourism sector during the year, accounting for 10.26 percent of the total employed population in the nation.[1] The term "ecotourism" has been widely adopted in China among government officials, research analysts, and tourism operators, as it demonstrates potential for meeting policy goals in local poverty alleviation and community development, while also fostering environmental and cultural understanding, appreciation, and conservation.[2] Meanwhile, interest is growing in wildlife tourism in China.[3] Nature-based tourism more broadly has developed rapidly since the 1990s and has been an important and distinctive part of the nation's tourism-industry growth strategy, written into the latest five-year plans of many provincial governments of inland regions rich with natural resources or unique ethnic heritage.[4] While it has been adopted in many areas as a solution to raise income level of rural regions without the harm of industrial pollution,[5] the mechanisms by which tourism is developed, operated, and organized often reflect complicated political, economic, and sociocultural conditions, including considerations that do not encourage the sustainability of healthy ecosystems. Ecotourism, therefore, is an important site for environmental communication research.

Wildlife tourism, in particular, aims to educate visitors about the threats facing wildlife and the need to protect particular species, including their habitats.[6] For example, popular interactive wildlife tours of the giant panda, a vulnerable species, are offered at the Chengdu Research Base of Giant Panda Breeding. If the experience is carefully designed, managed, and delivered, wildlife tourism has the potential to influence the conservation, knowledge, attitudes, and behaviors of tourists and other visitors.[7] Arguably, wildlife tourism has been essential in moving giant pandas from endangered status to vulnerable, and hopefully someday off any list of threatened species. Beyond specific species, Ballantyne et al. examined participants' memories of wildlife tourism experiences and found multiple channels through which such experiences could lead to long-term changes in conservation behavior, including sensory impressions, emotional affinity, and possibly reflective thoughts and behavioral responses.[8] Experts, however, recognize that the values of conservation, animal welfare, visitor satisfaction, and profitability are often in conflict in wildlife tourism—and tradeoffs in values and outcomes are common. How to balance and orchestrate the various elements of wildlife tourism practices in order to foster sustainable and positive connections with nature instead of harmful ones, therefore, is worthy of further study.

Wildlife tourism management choices also vary between anthropocentric and ecocentric worldviews, in turn reflecting different environmental ideologies, which are defined as "a way of thinking about the natural world that a person uses to justify actions toward it."[9] Guiding much wildlife tourism management is the anthropocentric assumption that humans are above and morally superior to nature, meaning that both nature and the nonhuman animals in it are viewed as objects to be managed for the pleasure of humanity without much consideration for the welfare of the nonhuman. On the contrary, an ecocentric worldview sees the world as an intrinsically dynamic, interconnected web of relations in which there are no dividing lines between the human and the nonhuman.[10] There is more recognition that animals also suffer pain like humans do, and that their rights and space should be respected. From this perspective, for example, close proximity to humans might be considered dangerous; touching by humans or flash photography might disturb or distress young animals if they are sensitive. In many parts of the world, wildlife tourism, as a subset of nature-based tourism and in its overlaps with ecotourism, seems to have the potential to accept a change in management strategies from an anthropocentric to an ecocentric ethic, thus further fostering sustainable aspirations.[11] Ideally, then, wildlife tourism can serve as an experiential trigger for imagining new forms of environmental communication.

In environmental communication, animals are often symbolic shorthand for environmental issues, representing pristine places in need of protection. In her study of orangutan tourism, for example, Stacey Sowards argues that humans can connect and identify with wild animals from an ecocentric perspective. She identifies a process that "allows humans to see the nonhuman world as a place in which we co-exist and as something other than an entity for human pleasure or exploitation."[12] Stephen Kellert also highlights the need for systematic investigation of human perceptions of nonhuman animals, as animals have influenced human modes of thought and belief since ancient times. From childhood, an individual develops affective, emotional relationships to animals; later there is a growth in cognitive, factual understanding and knowledge of animals, and eventually a "broadening of ethical concern and ecological appreciation of animals and the natural environment."[13] In a national sample of three thousand respondents in the United States, Kellert identifies eight value orientations toward animals.[14] The most common are humanistic (interest and strong affection for individual animals, especially pets), neutralistic/negativistic (avoidance due to indifference or dislike and fear), moralistic (concern for right/wrong treatment of animals and opposing cruelty/exploitation), and utilitarian. There are also naturalistic (interest and affection for wildlife and outdoors) as well as ecologistic (concern for environment as a system) value orientations. The least common are dominionistic (mastery and control of wildlife primarily as sport) and scientistic (interest in animal physical attributes and biological functions).[15] Researchers analyzing attitudes toward wolves in Norway using this typology found that a person's age, education, and level of experience with animals could affect their opinion on policy toward the species.[16] As Sowards's and Kellert's work indicates, communication *about* animals—and perhaps *with* them—is a promising area awaiting further studies.

Given the popularity of wildlife tourism and these emergent research findings of cross-cultural perspectives, environmental communication would do well to study wildlife tourism impacts. This essay accordingly focuses on Mount Emei, located in Sichuan province, which is one of the biggest pilgrimage centers of Chinese Buddhism, and one of UNESCO's World Heritage Sites. The booming tourism industry in China has made the geographically and culturally diverse regions of inland China more popular among domestic and international visitors, particularly since China initiated a multistage strategy in 2000 to boost the economic development of twelve western provincial-level regions, which are home to more than four hundred million people.[17] Mount Emei is of great spiritual and cultural importance, for it is one of the four holiest Buddhist mountains and one of the five holiest Taoist mountains in China. With an elevation of 10,000 feet and an area of 60 square miles, Mount Emei is attractive to travelers for its steep-sided terrain, gorgeous scenery, various microclimates, and diverse vegetation. Traditionally, travelers take the winding footpath, spending several days walking to the top to visit the Buddhist monasteries. With the rapid development of the region after the 1980s, however, the growing number of tourists and pilgrims that visit the mountain area are becoming a threat to the ecological system. According to UNESCO, a cable-car system imported from Austria in the 1990s leads to the Golden Summit of Mt. Emei, bringing more than 300,000 people a year to the mountain forest zone, where they can take a monorail connecting one summit to another.[18]

Visitors to Mount Emei will likely meet groups of Tibetan macaques, who are on the list of second-level protection in China. Travel agencies advertise the unique opportunities of interacting with the legendary monkeys, who are well known in folk tales for their spiritual connections with the Buddhist monks, while warning tourists that they can be very aggressive, sometimes grabbing food and opening bags. With the increasing amount of tourists to the region and the installation of the cable lines, the monkeys have changed their habits to stay closer to the tourist routes and many have become "fat and lazy," as described by locals to *China Daily*. Animal experts at Mount Emei Management Committee found that among the 1,200 monkeys in the region, many suffer from hypertension and diabetes.[19] Some tourists say that the monkeys have adopted a 9–5 schedule due to the open hours of the reserve. The tourism staff of the monkey reserve, armed with slingshots, monitor the activities of the monkey population by breaking up group fights among the animals, or keep them in a desired range, and sometimes help tourists locate cameras and phones snatched by them. Due to reported incidents of monkeys injuring tourists, a clinic has been established at the reserve to offer medical help. As these sources indicate, the boom in wildlife tourism at Mt. Emei has radically altered the local ecosystem.

In Chinese tradition, monkeys have been featured with human characteristics, as represented by the well-liked Monkey King figure in the sixteenth-century novel *Journey to the West*. In this classic tale, the Monkey King accompanied the Tang dynasty

Buddhist monk Xuanzang on his travels to Central Asia and India to obtain Buddhist sacred texts. The story of their adventure has been retold in many forms in school readings, comic cartoons, film, and TV dramas. The Buddhist tradition emphasizes a link between the human realm and that of animals, as various texts remind people that they were once animals and can be reborn as animals again. Confucianism also sees humans as having obligations to nonhuman animals, as well as compassion for their suffering, though it does not provide a view of their rights.[20] The evidence of anthropomorphism (attributing human characteristics or behavior to animals) in Chinese public culture is still present today. As social media has flourished over the years, the monkeys of Mt. Emei have become a popular topic in Chinese environmental public culture, for instance at the microblogging site Weibo. While many videos clips or photos focus on their mischievous acts, or amusing images holding a coke bottle or a map, sometimes the discussions have involved the policies of the Natural Ecological Monkey Reserve. In the winter of 2016, the Chinese zodiac Year of the Monkey, a photo of an injured wild monkey with hashtag #Monkey in Mount Emei seeking human help# received 3.75 million Internet users' attention with more than two thousand comments.[21] The photo showed a monkey with a nylon cord on its infected leg approaching a young tourist in the snow, attracting the attention of a local wildlife organization. The tourism administration staff spent ten hours in a temperature of minus 10 degrees Celsius to locate the monkey, treat the wound, and then release the twelve-year-old "monkey king" of a group of around forty monkeys back into the wild, to the relief of netizens.[22] As this instance illustrates, even while anthropomorphism leads to caricatured stereotypes about animals, it can also be a highly successful strategy for reaching viewers.

To analyze cross-cultural responses to the Mount Emei Monkey Reserve, this chapter compares and contrasts TripAdvisor reviews in Chinese and English. I will address two research questions: First, how do Chinese and English-speaking tourists evaluate their experiences at the monkey reserve on TripAdvisor differently? Second, what are the cognitive and affective components of their opinions, as a reflection of their concerns at the ecotourism site? The cognitive component is associated with awareness, referring to what tourists know or think they know about the things and events they see—for instance, information about the geography, history, or infrastructure of the region. The affective component evolves around people's expressions of personal feelings and emotions that the experience might evoke, such as pleasure, excitement, or annoyance.[23] In this way, I offer readers a comparative analysis of Western and Chinese responses to Mt. Emei, hence sharing a case study of how green communication functions within the larger national imaginaries that shape US-China relations.

Cross-Cultural Research on Nature Tourists

Tourists might bring a wide variety of motivations when participating in wildlife tourism, or they might make different decisions about whether it is an acceptable form of activity or not. In her book *Communicating Nature: How We Create and Understand Environmental Messages*, Julia Corbett questions how we should understand concepts like "environmental awareness" and "environmental concern," and how such awareness or concern develops over time. Corbett points out that "an environmental belief system inhabits each individual and informs her or him about where humans 'fit' in relation to the rest of the nonhuman world," impacting how the individual values things like California redwoods, wild creatures, and ecosystems. A person's childhood experiences, a sense of place, and historical and cultural contexts could affect the formation of an environmental belief system. When the environmental belief system is developed, it can be identified as an environmental ideology. Environmental ideologies could vary from *instrumentalism*, which focuses on immediate human desires, or *conservationism*, which emphasizes "wise use" of natural resources for the greatest number of people, or *preservationism*, which highlights the value of nature for scientific, ecological, aesthetic, and religious worth, to *radical transformative ideologies*, which seek to transform current anthropocentric relations to create more ecocentric relationships.[24]

In our information age, many would suggest that media coverage, public campaigns, environmental-education materials, and other forms of public communication can make an impact on people's perspectives toward the environment. Others might argue that personal experience, gender, and urban or rural settings make a difference. However, Ketil Skogen posits that such factors are not the only ones affecting attitudes regarding environmental problems and human relations to nature. Ketil argues that any communication activities aimed at influencing attitudes toward the environment need to note that such attitudes are part of the systems of interpretations and meaning that are embedded in cultural patterns.[25] From this perspective, environmental belief systems are both rhetorical and embodied, both cognitive and affective—and always open to the creative forces of interpretation and invention.

Travel in nature could ideally raise the public's awareness of the ecosystem and sensitize travelers to nature and its processes, achieving healthy outcomes for the body and mind. However, in practice, the motives of individuals might vary considerably. Traditionally, the Chinese notion of nature travel was influenced by perspectives on relationships between humans and nature rooted in Chinese literature, religion, and philosophy, resulting in the practice of interlinking cultural and natural attractions that may overlap with but also contradict Western values.[26] While in the West the aim or ideal for a protected area is often to mark off a pristine part of nature, shielded away from human impact, in China the aesthetic ideal may include human artifacts and art as part of the attraction for nature tourists, as sometimes seen in the romanticized verses by ancient adventure travel writers and poets.[27] Scholars also observe that, in contemporary China, the tourism industry operates under different assumptions than in the West; for instance, Buckley and coauthors note that many of the protected areas receive millions of visitors every year as part of the trend of "mass ecotourism," and that the domestic tourists often travel in significantly larger groups than their Western counterparts.[28] These Chinese practices might seem overwhelming to some Western travelers enjoying hiking alone in nature (in the style of Henry David Thoreau), though US tourists can show similar behaviors by swarming Yellowstone or Yosemite in the summer.

Still, research has found significant differences between Chinese mainland tourists and Western tourists in terms of their vacation preferences, expectations, and behaviors, while at the same time noting generational and gender differences.[29] Previous research shows that being close to or part of nature is a common theme in Chinese tourism, with both Confucian and Taoist teachings portraying travel as a means of learning, moral improvement, and enlightenment.[30] Chinese tourists are used to experiencing nature through associations with high culture elements such as landscape poetry, paintings, and calligraphy. Likewise, traditional popular culture, such as folklore, leads to practices such as praying to nature.[31] Scholars conducting a comparative analysis of feelings and attitudes toward nature among visitors to national parks in England and China found that an individual's inherited need for affiliation with nature is influenced by their sociocultural environment, in particular their national culture, but also by their current living place.[32] As these studies indicate, our affective relationship to the environment, and our experiences of how we interact with it hinge on a multitude of factors, including nationality, race, gender, age, class, and so on.

While the contemporary worldview of Chinese public culture in general is often characterized as anthropocentric or instrumental, there is growing recognition in Chinese society that environmental deterioration in urban areas is a serious concern. This shift in values has led to an increased desire, and even nostalgia in popular discourses in mainland China for, "green" mountain and clear water experiences of preindustrial purity. Consider, for example, the growth of *nongjiale* tourism, involving peasant families hosting urbanite guests in their guesthouses with the appeal of healthy rustic food and authentic, traditional lifestyle. White-collar residents in large cities flock to scenic areas during the weeklong holidays of May and October, while many of the wealthy fly abroad for vacations.

Earlier research surveys found that Western international ecotourists visiting China had higher environmental awareness and greater respect for traditional culture than their Chinese domestic counterparts.[33] Researchers also found that in southwestern and western provinces such as Guangxi, Yunnan, and Qinghai, both legal and illegal wildlife trade markets were flourishing, even though the government has been taking steps to strengthen wildlife management.[34] Based on a survey distributed in various trading places, published in 2008, scholars found that half of the respondents agreed that wildlife should be protected, yet 60 percent of them had consumed wildlife at some point in the last two years for food or traditional medicine.[35] Here, then, is another instance where respondents' stated beliefs and reported practices do not align, as a reflection of the complicated sources of influence on this issue in China. Meanwhile, among Chinese tourists traveling overseas, the picture is even more intricate. For example, a recent survey of Chinese and Australian visitors to a nature-based island resort in Queensland, Australia, found that Chinese visitors tended to have (1) a greater sense of connection with, but a more anthropocentric view of nature than Australian visitors; (2) less experience with, and a greater fear or dislike of animals in the wild; but (3) greater awareness of, interest in, and concern about environmental issues, such as the effects of global warming.[36]

TripAdvisor as a Research Tool

Social media plays a significant role in many aspects of tourism, including in the information search and decision making before a trip, in tourism promotion, in managing interactions with consumers during a trip, and in the circulation of vacation photographs and stories after a trip. The popularity of smartphones has enabled a vast amount of visitors to become agents for spreading electronic word of mouth, both positive and negative. The increasing popularity of tourist social media offers new possibilities to assess how people use and respond to nature and other cues for recreation and leisure travel. Many TripAdvisor reviewers post warnings and advice to potential future guests about a particular location they have already visited. Wasko and Faraj suggest that the process of exchanging knowledge based on experiences with the public could reflect an altruistic concern for others or a sense of public duty as a means to help others.[37] Most readers perceive user-generated travel reviews as being more likely to provide up-to-date, enjoyable, and reliable information in comparison to that from travel service providers.[38] Travel reviews posted in a personal capacity could potentially hold greater value, trust, and reliability than reflections collected by research professionals.[39] In short, examining the content of tourist-generated social media on the natural reserve yields rich data about tourist motivations, expectations, experiences, and levels of satisfaction about their travel destination. In this way, my data offers a snapshot of green communication in action.

TripAdvisor is the largest travel website in the world, with more than 315 million reviewers and over 500 million reviews published.[40] In April 2009, TripAdvisor launched its official site in mainland China, Daodao.com. In 2015 it was rebranded as Maotuying, a phonetic wordplay combining the Chinese characters for "owl" (the TripAdvisor emblem) with the Chinese character for "journey." In my analysis of Mount Emei Monkey Reserve TripAdvisor reviews, I focused on texts written in simplified Chinese language and English. As scholars using content analysis of online tourist websites have done before me, the main English-speaking countries are sorted to "Anglo," including the United States, Canada, Australia, and the UK.[41] Due to the fact that those who posted reviews were only a fraction of the travelers to the region, the reviews may not be representative of all tourists' opinions, and "false reviews" could sometimes appear for marketing purposes. With the increasing role of social media in tourist experiences, however, this method has been used for collecting valuable insights about traveler attitudes.[42]

Travel Review Research Findings

On the TripAdvisor website, under "Mt. Emei Natural Ecology Monkey Reserve," there is a total amount of 369 reviews, ranging from "excellent (5.0)" to "terrible

(1.0)." As this chapter goes to press, the average score of this tourist site is 4.0. Among all reviews, 45 percent gave a 5.0, and 40 percent gave a 4.0, and only 3 percent gave a score of 1.0 or 2.0. The earliest was posted in September 2010, and the latest review examined for the project was posted in October 2018. While a separate entry exists in TripAdvisor under "Mt. Emei" (Emeishan), with more reviews, it was not analyzed for this project as the topics discussed are too broad, mostly on routes, lodging, transportation, and cost, but less on environmental issues. Under the reviews for "Mt. Emei Natural Ecology Monkey Reserve," the reviewers also posted 118 photos, including scenic mountain views, temples, and monkeys of various sizes by themselves or near tourists. Some photos show tourists offering food to monkeys or posing with monkeys on their shoulders. One of the photos shows a big sign warning people not to approach or touch the monkeys, but apparently this was ignored by many tourists.

Visitors to Mt. Emei are of different nationalities, but the majority of reviewers are domestic tourists. The reviews include 43 written in English and 326 in Chinese (including 14 in traditional Chinese). There are also three in Russian, three in French, two each in Japanese, Italian, and German, one in Portuguese, and one in Spanish. Some reviews are very brief comments, while others are more extensive and might almost be classified as blogs. For the research project, the reviews written in languages other than English and Chinese were not selected.[43] Among the English reviewers, six indicated they are based in China, six did not offer a location, and the others self-identified as being from the following countries: four from Australia, three each from Canada and Hong Kong; two each from New Zealand, the United States, and the UK; seven from/based in Singapore; one each from Norway, the Netherlands, Romania, Malaysia, Indonesia, the Philippines, and Taiwan. As this list indicates, Mt. Emei is a truly global destination, attracting interest from around the world.

The data analysis takes the approach of identifying the affective, cognitive, and/or evaluative or action-recommendation content of the reviews among posts written in both languages. The following themes emerged: (1) observation of monkey activities; (2) feelings evoked by the presence of the monkeys, and thoughts and ideas on dealing with them safely/properly; (3) observation and evaluation of the tourism workers (including administration staff and local peddlers), which include their actions dealing with the monkeys and with the tourists; and (4) scenic environment. Also found but less frequently seen were (5) evaluation of other tourists; (6) perspectives on the monkey habitat, and feelings and views on ecology/natural resource management; (7) cultural/religious elements, in particular the Buddhist cultural artifacts and temples; and (8) notes on infrastructure and transportation. While the reviews in both languages have similarities in some of the themes, such as (1) observation of monkey activities, (4) scenic environment, (7) Buddhist culture, and (8) level of convenience of transportation, there are significant differences in (2), (6), and to some extent (3) and (5), as I will explain below.

Chinese TripAdvisor Reviews

Among the reviews in Chinese on the TripAdvisor site for "Mt. Emei Natural Ecology Monkey Reserve," 87 percent gave ratings of 4.0 or 5.0 to the monkey reserve area. The majority of Chinese tourists used positive terms to describe the characteristics of the monkeys, including "lovely," "active," "mischievous," "friendly," "fun," "bold," and "clever." Nearly a dozen tourists also noted the monkeys were "chubby," "fat," and "well-fed by tourists." A small portion of Chinese-language tourists used negative terms to describe the monkeys, calling them "bully," "reckless," "rampant," and "nasty." Many tourists indicated they were scared of the monkeys, but found it stimulating as they treasure the opportunity of getting close to wildlife and gave high ratings to the location. In terms of feelings evoked by the presence of the monkeys, many Chinese-language tourists showed feelings of affection for the monkeys, indicating that they are amazed by and appreciative of the fact that the monkeys of Mt. Emei are not fearful of humans and can coexist with humans "in harmony." For example, Xinlaidewawa from Shiyan city wrote: "Going to Mt. Emei, what made me happiest was the monkeys. It turns out to be true that they are not afraid of people. When we fed them, to our surprise and delight, one monkey actually grabbed the food bag from us. It was a really mischievous fellow." In this instance, the anthropocentric response indicates how visitors respond to the monkeys who have formed a habit of counting on the tourists' snacks as a food source, albeit in ways that might not be best for the monkeys' long-term health and environment.

The behaviors of the monkeys in the area also are widely discussed prior to many visitors' arrival, with cautionary tips from the guide that shaped tourist perceptions. For example, WeiyangNo1 from Sichuan province wrote: "We heard the monkeys in the area are divided into civilized monkeys and robbers, which we are afraid of. Fortunately, we did not provoke anyone along the way, nor did we encounter the legendary robber monkey." Another tourist, Meifazi from Shenyang, described the unique and exciting feeling of touching a monkey who came to search for food from his pockets, reminding him of the touch of a lover:

> The monkeys in Mount Emei are impressive and basically all ruthless . . . It's so interesting when the monkey reaches a hand into your pocket without saying anything, there is a bit of fear and curiosity; [it makes you] nervous as you can't get rid of him and can't help touching the monkey's hand (or I should call it paw). You feel your heart beat like touching a lover . . . Oh this is too much . . . Isn't it fascinating?

Reviews such as this appear to show confusion about the experience while trying to make sense of the unique interaction with wildlife. Comparing approaching the monkey to courting "a lover" again confirms my thesis that most of the responses to Mt. Emei are written in an anthropocentric manner that sees the animals through the lens of human values and experiences.

Some Chinese-language tourists noticed how the diet preferences of the monkeys have been influenced by the large quantity of tourists, such as this review written by Zhouningnan from Sichuan: "The monkeys there are now very picky, and they will not even give a look if you offer them ordinary corn kernels, but I think it will be interested in your Coca Cola and mineral water." Of course, the expectations of the visitors might turn out to be different from what they experienced, and some were disappointed for not seeing any wild monkeys themselves at the mountain. Finding out that the monkey groups moved away from the tourist area in cold winter, for example, one tourist ("Zhehuobushixiliya, China") expressed concern that in winter the monkeys do not have enough food to eat, as they saw few monkeys near the travel trail, and learned that the seasonal change of tourism affected the activities of the monkeys: "The staff said the monkeys went inside the mountains to find food in the winter, and now they don't come to the tourist area for food because there are few people." Information such as this appears to be shared to educate other tourists, as the reviewers acquired new knowledge about the habits of the monkeys.

While a portion of the Chinese-language reviewers noticed the body shape of the monkeys as being fat or obese, only one reviewer indicated it is not healthy, while many refer to it amusingly with the word "futai" (a positive term associating someone's weight with being rich). They appear aware that the monkeys' weight is a result of eating food from the tourists; however, I did not find a critical attitude toward the fact of monkeys eating human food or tourists handing out food to the monkeys. What brought out a negative feeling among some Chinese-language reviewers was that they noticed the monkeys are conditioned by the human contact, and it feels "unpleasant" and "against nature." A tourist, Sawyerli from Zhejiang province, wrote about the monkeys:

> They are really cute and funny, but they are too "snobbish." When seeing the passers-by, they will grab their thighs for food, and will not let them go unless they get food from them. A pack of monkey food [costs] 5 yuan, and just a little bit of corn. Without [this] human conditioning, this scenic spot will be closer to nature. I think we would like it more as tourists.

The commercial aspect of the monkey encounter—either photo-taking with trained monkeys or "bribing" monkeys to get off the bodies of tourists with so-called "monkey food" purchased from peddlers—gave some Chinese-language tourists negative feelings. For example, one post read: "They are not mountain elves, but have become a tool for workers to make money." Indeed, a considerable amount of the Chinese reviewers expressed critical opinions of the way the local people changed the nature of the monkeys and made them lose their innocence or naiveté, so they became a part of a profit-making scheme. For instance, Wanxi (no location) wrote: "The monkeys in Emei can really be called robbers, climbing onto you, and if you don't give them food, the person next to it, the robber's assistant, would come over saying, 'Buy a pack of monkey food, or it will not get off you. It might get angry and scratch or bite you.'"

One reviewer (Dandanegg from Shanghai) expressed disappointment over the influence of the humans on the wild animals, causing them to become part of a commercial activity and to lose their purity:

> The monkeys here lack the aura of the high mountains, but are more like those domesticated by humans and the monkeys in the zoo. Except for the little monkeys in their mother monkey's arms who have some innocent look in their eyes, most monkeys have a lot of worldly air in their eyes. I don't like this feeling very much.

This post indicates one of the central ironies of Mt. Emei, as a visitor has come to a well-traveled site, surrounded by all the amenities of the tourism industry such as a mountain lift and a clinic, yet expresses surprise that the monkeys lack a more pristine "aura." The tourist is implicated in a network of commercial exchange, yet is disappointed to find that "innocence" is missing.

While some tourists gave negative ratings to the location because of the aggressive behaviors of the animals, only a few tourists criticized the arrangement of the ecological reserve to allow close contact between humans and monkeys. Still, not all Chinese-language reviewers agreed with the policy of the monkey reserve to allow the monkeys to have a lot of freedom around humans. One female reviewer ("19810815" from Sichuan province) noted that the wild behaviors of the monkeys became more "rampant" as a result of protection:

> The result of overprotection is that monkeys are becoming more and more rampant, knowing that people will not hurt them. They will run to grab food, and there have been biting events (my aunt). It can't be said that it's bad, it's just their wild nature.

Overall, then, we find a strong tendency to perceive the monkeys in anthropocentric terms, even while holding onto a romantic view of nature.

English TripAdvisor Reviews

Reviewers writing in English tend to write longer reviews than the Chinese reviewers. Around half of the reviews in English rated the monkey reserve "average (3.0)" or "poor (2.0)" and "terrible (1.0)." Notably, this is lower than the average scores given by the Chinese tourists. As noted above, the background location of these reviewers varied, but I will analyze these 43 reviews as a heterogeneous cultural contrast to Chinese-language reviews. The majority of reviews in English commented on the monkeys, with observations of their behaviors, their own experience encountering them, and opinions on the living environment of monkeys at the reserve. While some were surprised by the monkeys, most had learned cautionary information beforehand from tour guides or books and appeared more prepared, to various extents, for the experience. The English-speaking reviewers used both positive and negative words to describe the monkeys, ranging from "cute," "adorable," "cheeky," "funny and cool" to "super active," "naughty," "aggressive," and "scary." Some also described the shape, size, and even state of health of them, using words such as "little," "heavy," "active," or "obese." The reviewers also used a variety of verbs to depict the activities of the monkeys, including "jump," "rob," "grabbed my thighs," "unzip bags," "check your pockets," "climb on you," "sat on shoulders," "scratch," and "steal." The behaviors observed are similar to those written in Chinese. Like their Chinese counterparts, a few tourists attributed human characteristics to the monkeys in an amused manner; for instance, a male tourist from Singapore jokingly described the monkeys and the peddler as "bandits," writing, "An obese monkey grabbed my thighs saying FEED ME."

Even though many English-language tourists were well informed prior to the trip, the sight of wild animals could bring out a lot of unexpected raw emotions from the tourists hiking a narrow path. One traveler from New Hampshire, using the handle of "meepbeepeep," wrote about the "unbridled fear" caused by the encounter with the wild monkeys "the size of pitbulls" who could "dangle over your head," and offered lengthy advice to TripAdvisor readers with five main points, including not wearing bright colors and not putting hands in pockets. Here is an excerpt:

> #1. GET A BIG, STURDY STICK. Sometimes, the bastards come at you. You keep that stick because while most of the macaques are small-ish, there are some huge honking mamas who I'm pretty sure could have wrecked me. If they come after you (and I saw some tourists have monkeys go after them), you'll be glad of the stick. No, I don't support hurting animals, but these are not super fragile and dainty animals.

This reviewer acknowledges that the monkeys are "interesting" and "impressive," and offered five tips to go through the monkey area safely, giving a score of 4.0 to the location.

Another reviewer, "livo815" (no location), gave a rating of 4.0 and wrote about the ironic fact that humans willingly take risks of being injured to get close to the naughty monkeys, despite the payment asked and the line in front of the medical room waiting for treatment:

> By the time you get to the "monkey zone" the crowd is so thick it's like standing in line, garbage is strewn everywhere, naughty monkeys swat at children, attack adults who failed to hide plastic bottles, the line to the medical treatment booth was 10 deep, yet still, happy tourists continued to pay 5rmb for little packets of sunflower seeds and dried corn. We opted to not feed the monkeys, but still ended up with a very aggressive female literally on my back. Tried to use my walking stick to make a noise to scare her and she just grabbed onto the stick and tried to yank it away. Alas. I loved every moment of it despite being terrified.

Likewise, a male tourist from San Diego, California (Jack Y) described the monkeys as "scary" but gave 5 stars to the location, indicating that over 99 percent of the people got out without any incidents, and praised the beautiful path. The tourist also paid to have his two kids take pictures with a monkey under the supervision of a ranger:

> I saw one of them jumping onto the shoulder of a tourist. The tourist, out of panic, grabbed the monkey and threw it to the river. The monkey grabbed the bridge and did not fall into the water. But the monkey was really angry and made all kinds of noise to the poor guy. It looked like he was going to attack the tourist but eventually backed off.

It seems that any moment at this site could be action-packed with thrilling and risky encounters between the humans and the animals, making the experience tense and memorable. Perhaps this gives the tourists the anticipated contrast with their routine lives, making their long journeys worthwhile and motivating them to write extensive online reviews.

The reviews in English have rich cognitive content on the relationship between tourists and wild animals, and their ideas of what is a "natural" lifestyle for monkeys. One reviewer noted with indignation that the monkeys attacked only the tourists but not the locals, probably because the locals beat them with sticks. Many of the travelers expressed their opinion about human-monkey relationships, stating that humans should not give monkeys food that is unhealthy for them, or make the monkeys aggressive by provoking them. A couple from Australia identifying as "John and Deb K" indicated that only natural food from the forest is proper for them:

> These monkeys are getting a bad deal—unfortunately tourists are feeding them— this causes them to behave badly. It is wonderful to see them but not what is occurring—humans feeding them food that is not good for them but is full of sugars so yes they get addicted just like people to unhealthy foods. Please don't feed them and if you have food put it away and hold on to your bags.

This couple not only is concerned about the nutritional health of the animals, but also is making recommendations to readers of TripAdvisor.

Some reviews in English showed sympathy for the monkeys and criticism of the tourists and the way they treat the monkeys for their entertainment. A review by "gibsonwushu," a male tourist from Oslo, Norway, is concerned that the mental state of the monkeys has been affected by humans: "These monkeys were just fantastic, but unfortunately they get stressed out by the disrespect many of the tourists show them. They would have been lovely kind animals if everyone let them be just that and not treat them as dolls for their own entertainment." Likewise, "Ian P," a male tourist from The Hague, Netherlands, said he felt "very sad" about the way the monkeys get used to the mainland Chinese tourists. He also made evaluative statements about the actions of other tourists:

> Stupid mainland tourists have corrupted these monkeys to become used to humans. They are very aggressive now and will come to you to try and steal food. They have learned to grab plastic bags and such because they know food is in there. I passed this area quickly because seeing the tourists dealing with these wild animals the way they did despite the clear signs warning them of the danger made me very sad.

Further, "Annarita" (a young woman based in China) criticized the way the local people use the monkeys for the purpose of profit-making, and she points out that this should not be allowed at a natural reserve. Apparently, she has a clear notion of what a "real" natural reserve should look like:

> I've been harassed by three monkeys and luckily they only stole a spring for hairs. I even saw a man who confined some monkeys on a bridge, you can pay him and get a picture with a monkey on your shoulder! What a shame! Those animals are supposed to be free! It is supposed to be a monkey reserve!

Some tourists reminded others of the health risks to humans at the monkey area. One tourist, "krappkrapp," who identifies as a *Lonely Planet* reader, warned other tourists of the danger of a monkey bite and recommended hospitals in the nearby city of Chengdu, where someone could understand English:

> The landscape in this area is very beautiful and the monkeys are funny and cool— BUT: if you love your life, don't go there without having an anti-rabies inoculation before! . . . Don't be afraid of Chinese hospitals, they are OK. Try to find out the Chinese word for rabies injection. I also showed the doctor a picture of the monkey and he understood my problem.

Likewise, "Pat C," a male tourist from Singapore, criticized the local "handlers," commenting that it is "neither safe or hygienic" to let monkeys get on people's shoulders as they might have fleas; expressed concerns over the diet of the monkeys; and indicated that tourists have responsibilities to help them eat healthy food:

> The adult macaques are obese and gobbling chips and crisps . . . Yet, there are signs saying not to feed the monkeys. It is unforgivable to introduce processed snacks, sugared, fried in oil and salted, to the monkeys' diet. Some look bloated and puffed up. If these "handlers" must cash in on the show, perhaps they should sell fruits that the monkeys feed on . . . Let's act to preserve this specie of primates. Bring your own fruit when you visit.

In that same vein of commentary, some tourists writing in English made strong comments on the lifestyle of the monkeys in the area, naming specific possible diseases; for instance, "Eternal_Nomad-12," a visitor from London, expressed negative feelings about the local vendors, and concern for the health of the monkeys eating human food:

> The way it works: Monkey brazenly steals your valuables. Vendors suggest you buy food from them to get it back. You proceed to buy the food in exchange for your valuables. Repeat the process for the next victim. What a scam! . . . Do NOT feed the monkeys—at all. I have already seen enough obese monkeys, having subsisted on our fatty snacks, that I wouldn't be surprised if they already have diabetes or heart disease.

A male tourist from Australia, "monkey2009," who gave the location a low score of 1, criticized the arrangement of the reserve harshly and asked other tourists not to go there:

> This place is a disgrace . . . There is nothing Ecological about this circus show and they should divert the trek to avoid disrupting the monkeys from their normal lives and seeking human intervention for sustenance (crackers are not their natural food) . . . Do not go there and do not bring your children there!

This is a clear evaluative comment with an action recommendation, highly critical of the lack of ecological concern in the arrangement. By calling Mt. Emei a "circus," "monkey2009" implies the venue is not so much an immersion in environmental thinking as an exercise in exploitation.

Overall, around 45 percent of the tourist reviews in English rated the monkey reserve "average" (3.0) or lower. The major concerns were the safety and health risks to humans due to the monkeys' aggressive behaviors, the unhealthy diet the monkeys have by eating human food, the contaminating presence of tourist crowds that took away the natural habitat of the animals, causing stress, agitation, and aggressiveness by the monkeys toward humans. Some of the reviews complained about the profit-making motivation of the local monkey handlers/peddlers selling monkey food or bamboo sticks who are not always trying to help. Negative words such as "sad," "disturbing," and "terrible" were used to comment on the ecological environment of the monkey reserve area. The average level of enjoyment among reviewers writing in English seems to be lower than their counterparts writing in Chinese, as they are more cognizant of the unseen risks to themselves of human-monkey contact beyond getting their bags opened or their skins scratched. Indeed, they were uncomfortable with the uneasy reliance of the monkeys on the tourists and were troubled by the ecological implications of such an arrangement. Their expectations of an "ecological natural reserve" might be very different from those of the large crowds of Chinese tourists and busy commercial activities along the paths, leading to lower ratings, even though many expressed amazement over the scenic views and the charm of the monkeys.

Conclusion

While environmental ideologies of preservation and conservation can be found in reviews in both languages, there are far more evaluative and behavior-recommendation reviews in English, many including ecological considerations. Affectively, there are obvious similarities among portions of the Chinese and English-speaking reviewers, who shared the sense of scariness or excitement and fun when encountering the monkeys face to face. Cognitively, there is a clear pattern that contrasts the majority of the English-language TripAdvisor reviewers with their Chinese-language counterparts. A lot of the Chinese tourists showed a sense of curiosity and amazement about the wild animals, as if they were at a petting zoo, and enjoyed the rare excitement of seeing them take food from the tourists or taking photos with them. Some also showed interest in biological information, such as monkey group activities, life expectancy, etc. In contrast, a significant number of Western reviewers were disturbed by this "petting" practice and held evaluative and critical attitudes about other tourists, as well as the tourism administration. This could be partly due to the fact that many of the English-speaking tourists to Mt. Emei were seasoned travelers with familiarity with ecological knowledge and practices of natural reserves in developed countries; in comparison, the Chinese reviewers were mostly domestic travelers, who appeared to be young people who accessed the TripAdvisor website at its early-development stage in China. This difference could be seen to a certain extent from the TripAdvisor reviewer levels and "badges" of the individual profiles in both languages, though there is not enough information on customer profile characteristics to compare all of them.

Another reason that might lead to the differences could be the level of environmental education in China versus English-speaking countries as well as its highlighted features. For a lot of Chinese, the purposes of a natural reserve have not been well defined, except for protecting the safety of endangered animals such as pandas, often seen in the media. There is a lack of the notion that the areas should be kept intact away from human activities or presence. While there are occasionally criticisms against overdevelopment of the scenic area, referring to the amount of commercial activities and infrastructure construction, it seems to come from mainly an aesthetic perspective, as if a scroll depicting a beautiful green view of nature for the visitor is tainted by the footprint of modernity.

The understanding of ethical and responsible wildlife tourism has become more complex with the growth of online information sources and social media platforms in countries like China. Sometimes "good" intentions to help wildlife may not be the most ecologically sound solutions. Environmental awareness education through social media platforms could potentially provide individuals with the knowledge, skills, and attitude required to build a culture of sustainability. How the natural reserve administration addresses and responds to the issues raised in the social media content could be meaningful for managing the goals of natural-resources preservation and tourism development.

Furthermore, as the reviews studied for the project span across different years, the specific conditions of the monkey population might have changed to some extent. There are indeed a few reviewers in 2018 who mentioned that the tourism management agency has taken steps to ensure that both monkeys and humans behave properly to minimize incidents of injury, stating that the stories of aggressive monkeys are exaggerated.

For environmental communication scholars, cross-cultural comparisons of human-nature interactions help reveal assumptions and norms that are connected with deep-rooted belief systems and ideologies, as well as opportunities for transformative experiences. Contemporary tourist administrators tend to promote and brand wildlife sites as having conservational and educational significance, though critics can argue that many visitors come for entertainment and recreational purposes. The highly interactive tour arrangements could lead to questions of animal welfare and could also affect the ecologically sustainable outcome of the ecological area. Indeed, I found in this study that while most Chinese-language reviewers tended to reflect an anthropocentric view instead of the ecocentric view of English-language reviewers, they also expressed a sense of awe and admiration for the monkeys. Although many Chinese tourists came for relaxation and recreation, they were aware of the value of the mountain as the home of the animals and appreciated the brief, direct contact experience with the natural world.

Overall, wildlife tourism globally and in China is a significant site for negotiating environmental communication with possibilities for positive and negative reactions. This study has shown how TripAdvisor is one rich social media platform for analyzing tourist reviews across cultural perspectives. By focusing on the Mount Emei Monkey Reserve, we can draw comparative cultural observations of human expectations and opportunities to be transformed by wildlife encounters. Promoting greater care for wildlife and nature more broadly will require that we take seriously such social media reviews. We might speculate, then, that US-China relations could be enriched by building more intentional ecological perspectives into such tourism sites, which, ideally, could offer shared experiences in cross-cultural understanding and new green imaginations.

NOTES

1. "China's Tourism Contributes More Than 10% to GDP and Employment," *China Travel News*, November 21, 2017, https://www.chinatravelnews.com/article/118697.

2. John Donaldson, "Tourism, Development and Poverty Reduction in Guizhou and Yunnan," *China Quarterly* 190 (2007): 333–351.

3. Li Cong, David Newsome, Bihu Wu, and Alastair M. Morrison, "Wildlife Tourism in China: A Review of the Chinese Research Literature," *Current Issues in Tourism* 20, no. 11 (2017): 1116–1139.

4. Chunyu Liu, Junsheng Li, and Peter Pechacek, "Current Trends of Ecotourism in China's

Nature Reserves: A Review of the Chinese Literature," *Tourism Management Perspectives* 7 (2013): 16–24.

5. Jundan Zhang, "Tourism and Environmental Subjectivities in the Anthropocene: Observations from Niru Village, Southwest China," *Journal of Sustainable Tourism* 27, no. 4 (2018).

6. Jerry F. Luebke, Jason V. Watters, Jan Packer, Lance J. Miller, and David M. Powell, "Zoo Visitors' Affective Responses to Observing Animal Behaviors," *Visitor Studies* 19, no. 1 (2016): 60–76.

7. Roy Ballantyne, Jan Packer, and Karen Hughes, "Tourists' Support for Conservation Messages and Sustainable Management Practices in Wildlife Tourism Experiences," *Tourism Management* 30, no. 5 (2009): 658–664.

8. Roy Ballantyne, Jan Packer, and Lucy A. Sutherland, "Visitors' Memories of Wildlife Tourism: Implications for the Design of Powerful Interpretive Experiences," *Tourism Management* 3, no. 4 (2011): 770–779.

9. Julia Corbett, *Communicating Nature: How We Create and Understand Environmental Messages* (Washington, DC: Island Press, 2006), 13.

10. Ibid.

11. Ralf C. Buckley, *Ecotourism Principles and Practices* (Wallingford, UK: CABI Publishing, 2009).

12. Stacey K. Sowards, "Identification through Orangutans: Destabilizing the Nature/Culture Dualism," *Ethics & the Environment* 11, no. 2 (2006): 45–61, 59.

13. Stephen R. Kellert, "Attitudes toward Animals: Age-Related Development among Children," *Journal of Environmental Education* 16, no. 3 (1985): 29–39.

14. Stephen R. Kellert, "Affective, Cognitive, and Evaluative Perceptions of Animals," in *Behavior and the Natural Environment*, vol. 6 of *Human Behavior and Environment: Advances in Theory and Research*, ed. I. Altman and J. F. Wohlwill (Boston, MA: Springer, 1983).

15. Kellert, "Attitudes toward Animals."

16. Tore Bjerke, Ole Reitan, and Stephen R. Kellert, "Attitudes toward Wolves in Southeastern Norway," *Society & Natural Resources* 11, no. 2 (1998): 169–178.

17. These regions are Chongqing, Sichuan, Guizhou, Yunnan, Tibet, Shaanxi, Gansu, Ningxia, Xinjiang, Inner Mongolia, Guangxi, and Qinghai. See China State Council, "New Five-Year Plan Brings Hope to China's West," http://english.gov.cn/premier/news/2016/12/27/content_281475526349906.htm.

18. UNESCO, "Mount Emei Scenic Area, Including Leshan Giant Buddha Scenic Area" (2018), https://whc.unesco.org/en/list/779.

19. "A Diet Fit for a Monkey King," *China Daily*, January 9, 2007, http://www.china.org.cn/archive/2007-01/23/content_1197045.htm.

20. Neil Dalal and Chloë Taylor, *Asian Perspectives on Animal Ethics: Rethinking the Nonhuman* (New York: Routledge, 2014).

21. "Netizens Rally Round Injured Monkey King ahead of Year of the Monkey," *China Daily*, February 1, 2016, http://www.chinadaily.com.cn/china/2016-02/01/content_23342956.htm.

22. "Injured 'Monkey King' Back in the Wild after Rescue," *China Daily*, February 4, 2016, http://www.chinadaily.com.cn/china/2016-02/04/content_23398994_2.htm.

23. Stella Kladou and Eleni Mavragani, "Assessing Destination Image: An Online Marketing Approach and the Case of TripAdvisor," *Journal of Destination Marketing & Management* 4, no. 3 (2015): 187–193.

24. Corbett, *Communicating Nature*, 12.

25. Ketil Skogen, "Another Look at Culture and Nature: How Culture Patterns Influence Environmental Orientation among Norwegian Youth," *Acta Sociologica* 42, no. 3 (1999): 223–239.

26. Cui Qingming, Xu Honggang, and Geoffrey Wall, "A Cultural Perspective on Wildlife Tourism in China," *Tourism Recreation Research* 37, no. 1 (2012): 27–36.

27. R. Buckley, Carl Cater, Zhong Linsheng, and Chen Tian, "SHENGTAI LUYOU: Cross-Cultural Comparison in Ecotourism," *Annals of Tourism Research*, 35, no. 1 (2008): 47–61. See also Ye Wen and Xue Ximing, "The Differences in Ecotourism between China and the West," *Current Issues in Tourism* 11, no. 6 (2008): 567–586.

28. Ibid.

29. Ruomei Feng, Alastair Morrison, and Joseph Ismail, "East versus West: A Comparison of Online Destination Marketing in China and the USA," *Journal of Vacation Marketing* 10, no. 1, (2004): 43–56.

30. Honggang Xu, Peiyi Ding, and Jan Packer, "Tourism Research in China: Understanding the Unique Cultural Contexts and Complexities," *Current Issues in Tourism* 11, no. 6 (2008): 473–491. See also Paul Harris, "Green or Brown? Environmental Attitudes and Governance in Greater China," *Nature and Culture* 3, no. 2 (2008): 151–182.

31. Qingming Cui, Xiaohui Liao, and Honggang Xu, "Tourist Experience of Nature in Contemporary China: A Cultural Divergence Approach," *Journal of Tourism and Cultural Change* 15, no. 3 (2017): 248–264.

32. Dorothy Fox and Feifei Xu, "Evolutionary and Socio-cultural Influences on Feelings and Attitudes towards Nature: A Cross-Cultural Study," *Asia Pacific Journal of Tourism Research* 22, no. 2 (2017): 187–199.

33. Buckley et al., "SHENGTAI LUYOU." See also Han Jia, Andrea Appolloni, and Wang Yunqi, "Green Travel: Exploring the Characteristics and Behavior Transformation of Urban Residents in China," *Sustainability* 9 (2017): 1043–1057.

34. Richard Harris, *Wildlife Conservation in China: Preserving the Habitat of China's Wild West* (New York: Routledge, 2008).

35. Li Zhang, Ning Hua, and Shan Sun, "Wildlife Trade, Consumption and Conservation Awareness in Southwest China," *Biodiversity and Conservation* 17, no. 6 (June 2008): 1493–1516.

36. Jan Packer, Roy Ballantyne, and Karen Hughes, "Chinese and Australian Tourists' Attitudes to Nature, Animals and Environmental Issues: Implications for the Design of Nature-Based Tourism Experiences," *Tourism Management* 44 (2014): 101–107.

37. Molly McLure Wasko and Samer Faraj, "'It Is What One Does': Why People Participate and Help Others in Electronic Communities of Practice," *Journal of Strategic Information Systems* 9, no. 2–3 (2000): 155–173.

38. Hee "Andy" Lee, Rob Law, and Jamie Murphy, "Helpful Reviewers in TripAdvisor, an Online Travel Community," *Journal of Travel & Tourism Marketing* 28, no. 7 (2011): 675–688.

39. Ingrid Jeacle and Chris Carter, "In TripAdvisor We Trust: Rankings, Calculative Regimes and Abstract Systems," *Accounting Organizations and Society* 36 (2011): 293–309.

40. TripAdvisor, "Fact Sheet" (2013), https://tripadvisor.mediaroom.com/us.

41. Liang Tang, Soojin Choi, Alastair M. Morrison, and Xinran Y. Lehto, "The Many Faces of Macau: A Correspondence Analysis of the Images Communicated by Online Tourism Information Sources in English and Chinese," *Journal of Vacation Marketing* 15, no. 1 (2009): 79–94.

42. Liping A. Cai, William C. Gartner, and Ana María Munar, eds., *Tourism Branding: Communities in Action* (Bingley, UK: Emerald, 2009). See also Ana María Munar, "Tourist-Created Content: Rethinking Destination Branding," *International Journal of Culture, Tourism and Hospitality Research* 5, no. 3 (2011): 291–305.

43. The Chinese reviews are translated by the author.

From Green Peacock to Blue Sky

How ENGOs Foster Care through New Media in Recent China

Jingfang Liu and Jian Lu

Since the 1990s, when environmental nongovernmental organizations (ENGOs) made their debut in China, environmental activism has grown vigorously in China. Along the same time period, the Internet and new media greatly transformed China. After almost three decades have passed, what has changed in regard to new media-based environmental activism in China? And how has the recent development of new media shaped the repertoire and dynamics of ENGO activism? To address these questions, this chapter investigates the role of new media in shaping the dynamics of recent environmental activism in China. Focusing on three case studies, we examine recent trends in how Chinese ENGOs strategically utilize new media to promote environmental and social change. Based on interviews with ENGOs, online ethnography, and documentary research, we present three cases: the Saving the Green Peacock campaign, the case of Lei Ping, and Institute of Public and Environmental Affairs (IPE)'s Blue Sky online app. Through these exemplars, we identify three notable mechanisms by which ENGOs use new media to perform their activism, namely: (1) making culture and emotion-based appeals, (2) flooding online sites with public-opinion-based calls to action, and (3) generating public awareness by offering science-focused information platforms. Taken as a whole, our three case studies demonstrate that while ENGOs often mobilize around crises (over an endangered species, the silencing of an activist, or pollution impacts on public health), their networks and impact also speak to a growing ethic of care for the environment in China.

China's ENGOs and Environmental Activism

ENGOs have been one of the key players in China's environmental activism since the early 1990s. In 1992, when Friends of Nature (FON) was established in Beijing as China's first grassroots ENGO, China's green civil society was born. Despite sociopolitical restraints, ENGOs have managed to emerge as an important component of civil society in China,[1] and have grown at an annual rate of 10–15 percent "in terms of both number of organizations and number of staff working for them."[2] Thus, according to the *Statistical Bulletin of Social Service Development*, issued by China's Ministry of Civil Affairs, there were 6,000 environmental social organizations (*Shehui Tuanti*) and 444 environmental private nongovernmental enterprises (*Minfei Qiye Danwei*) by the end of 2016.[3]

Along with the rapid increase of ENGOs, the past two-and-a-half decades have witnessed a vibrant, growing "green movement" of environmental groups with diverse formats, working on a wide variety of issues. In the early stage, Chinese ENGOs were mainly "engaged in environmental education, nature conservation, species protection, policy advocacy and many other activities" that are relatively less politically sensitive.[4] In recent years, however, the abilities of ENGOs in terms of assisting the public's defense of environmental rights, monitoring environmental policy implementation, and promoting enterprises' environmental protection responsibilities have grown stronger, especially in influencing environmental policy.[5] Chinese ENGOs also actively participated in the course of legislation that led to the new Environmental Protection Laws.[6] An exemplar of how ENGOs have pushed for policy changes in China is the nearly one-and-a-half-decade-long campaign against the construction of a hydropower station on the Nu River in Yunnan province, involving ENGOs, intellectuals, and media experts successfully pressuring authorities to shelve the plan of dam construction on Nu River in 2016.[7]

Still, the relationships between ENGOs and the state have been complex and dynamic. In order to survive under an (semi-)authoritarian regime, where the authorities are highly sensitive to confrontational activities, Chinese ENGOs have consciously focused on nonconfrontational activities. Peter Ho referred to this

ENGO-led environmental activism as "greening without conflicts."[8] He suggests that by preemptively "greening" the state, China prevents environmental values from evolving into a confrontational environmental social movement or broader public protests, which could challenge the legitimacy and authority of the state. On the one hand, the "greening" of the Chinese state, which refers to the state's ongoing efforts to reduce environmental pollution, has created some institutional space for ENGOs to engage in growing environmentalism. In China, most ENGOs choose to restrict themselves to nonconfrontational or depoliticized activities rather than more provocative political actions, such as protests, although some organizations participate in relatively confrontational activities (such as the anti-dam campaigns in Yunnan). Yangzi Sima has accordingly observed that "the non-antagonistic culture of Chinese civic ENGOs hence presents a considerable contrast to environmental activism in Western countries, where activists regularly engage in direct confrontations with authorities."[9]

Notably, the Chinese government has been an increasingly involved player in China's green movement. For instance, in trying to balance environmental protection and economic development, President Xi stated: "We need not only lucid waters and lush mountains (the environment), but also golden and silver mountains (the economy). We'd rather have lucid waters and lush mountains over golden and silver mountains. And lucid waters and lush mountains are golden and silver mountains."[10] In addition to such discourse, numerous environmental laws have been established since 2013 regarding water, soil, and air. Most noticeable is the New Environmental Law that came into effect on January 1, 2015. Regarded as "the strictest environmental protection law in history," this law introduces innovative provisions, such as successive penalties for law-breaking behaviors.[11] The State Council also issued "Ten Measures for the Prevention and Control of Air Pollution," "Ten Measures for the Prevention and Control of Water Pollution," and "Ten Measures for the Prevention and Control of Soil Pollution," respectively in 2013, 2015, and 2016.[12] In 2016, the "river chief system" (He Zhang zhi) was created and implemented nationwide.[13] Meanwhile, the central environmental-protection inspection team, known as the "imperial messenger of environmental protection," was set up to inspect provincial governments and party committees in terms of the performance of environmental protection.[14] These landmark changes—like the watershed moment of 1970 in the United States, when the US Environmental Protection Agency (EPA) was established and many major environmental laws were institutionalized—indicates that the Chinese government is willing to strengthen environmental governance, providing more opportunities for China's civil society to participate in environmental protection.

New Media and Social Activism

In the past two decades, China has experienced a rapid mediatization of communication technologies and practices. By 2017, China's population of Web users has reached 772 million, close to the sum of India's (462 million) and the United States' (312 million) combined.[15] As of December 2017, the Internet penetration rate in China reached 55.8 percent, exceeding the global average (51.7 percent) by 4.1 and the Asian average (46.7 percent) by 9.1.[16] With rapid development of information communication technologies (ICTs), the Internet has transitioned from the Web 1.0 to the Web 2.0 era, which has transformed ordinary individuals from passive receivers to active producers of online information.[17] Although "the Internet firewall" prevents Chinese users from accessing popular US-based social media platforms, such as Facebook, Twitter, and Instagram, China's own social media applications, such as Weibo (a microblog) and WeChat (a messaging app), are thriving. WeChat, a free social media application launched by Tencent in January 2011, has rapidly developed into a comprehensive, all-inclusive multimedia platform that integrates various functions such as social networking, marketing, and online payment. As of 2018, its monthly user size has exceeded 1 billion.[18] These new media tools allow Chinese people to connect, work, date, shop, and order take-out 24/7 at superfast speed, gradually creating a "Chinese society online." The rise and rapid development of new media are profoundly changing and shaping not only the ways information is communicated but also various aspects of social life. Chinese ENGOs have also kept pace with this development, and hence have been at the forefront of what Patrick Shaou-Whea Dodge has called the "cat-and-mouse game" of online activism, wherein authorities seeking to control discourse and activists seeking change dance around each other, creating an online culture of repression *and* subversion, both control *and* fluidity.[19]

From the classical indigenous Zapatista movement to the recent Occupy movements and People's Climate Marches, social and environmental movements worldwide have actively engaged new media to pursue social change.[20] Activists and activist groups have used new media to produce and disseminate information, coordinate online and offline collective actions, and exert influence.[21] Particularly, they have used various new media to engage in new modes of collective action, to build looser and flexible organizing structures that are networked, decentralized, and nonhierarchical, in contrast to traditional hierarchical organizations.[22] These structures facilitate horizontal communication and lead to new forms of instantaneous online mobilization networks.[23] Scholars from a wide range of disciplines are working to understand these changes, particularly in terms of mobilization structures, opportunity structures, and framing processes.[24] For example, Habibul Haque Khondker studied new media in the Arab Spring of the Middle East and North Africa and argues that new media such as Facebook played a critical role, especially in light of the absence of an open media and a civil society.[25] In examining the 2011 Egyptian uprising, Merlyna Lim finds that the most successful social movements were those using social media to expand networks of disaffected Egyptians, broker relations between activists, and globalize the resources and reach of opposition leaders. Social media provided the opposition leaders with the means to frame the issues, shape the repertoires, enhance collective identity, and transform online contention into offline collective actions.[26] We should note, of course, that these online organizing efforts were mapped onto long-standing union affiliations, student clubs, religious organizations, and other "offline" forms of political life.

In China, Yongnian Zheng and Guoguang Wu explored the role of the Internet in mobilizing collective actions and found that it does this through facilitating the free flow of information and providing alternative information sources.[27] Chinese ENGOs have engaged new media to voice their opinions, recruit volunteers, mobilize capital, and carry out various types of advocacy campaigns.[28] Given the stricter political environment, the difficulty of formal registration, and the lack of funding, Chinese ENGOs rely heavily on new media.[29] In the early stages, burgeoning Chinese ENGOs embraced the Internet for their cause. They built their own websites and email lists, hosted online forums or bulletin board systems (BBS), issued newsletters online, and used the Web to disseminate information, publicize their work, recruit, educate, organize, promote discussion and debate, and network with peers and international organizations.[30] Guobin Yang noted the rise of online activism in China and observed two types of online activism: "internet-assisted contention," for which the Internet is used as a tool to mobilize offline collective actions, and "internet contention," which refers to the popular contention that takes place in cyberspace.[31] One of us (Liu) studied twenty ENGOs and their Internet-based environmental collective actions, finding that the Internet combined with other conditions, such as a large alliance of organizations involved, has a great capacity to promote social change through environmental Internet activism in China.[32] Through a case study of a prominent ENGO, the Global Village of Beijing, Yangzi Sima also found that new media effectively empowered resource-poor activists in their self-representation, information brokering, network building, public mobilization, and construction of discourse communities.[33]

In the Web 2.0 era, as Chinese society becomes deeply entrenched in the media world in every aspect, Chinese ENGOs continue to be pioneer users of many new media platforms such as Weibo and WeChat. As soon as Sina Weibo came into service in August 2009, FON created their Weibo account and became one of the earliest ENGO users of Weibo. As early as October 2013, FON also created their official WeChat account, a highlighted example of Web 2.0 application. As the second generation of the Web, Web 2.0 enables "interoperable, user-centered web applications and services" that "promote social connectedness, media and information sharing, user-created content, and collaboration among individuals and organizations."[34] As Fengshi Wu and Yang Shen summarized, Web 2.0 applications bear a number of prominent features such as more information format, more mechanisms for public monitoring, and a higher level of transparency. These features allow NGO activists to employ various strategies to enhance public engagement, including monetary transaction, resource allocation, public monitoring, and large-scale interorganizational coordination.[35] Based on three Mexican social and political movements that span close to twenty years, Rodrigo Sandoval-Almazan and J. Ramon Gil-Garcia examined similarities and differences in the use of information technologies and propose a framework to understand the evolution of cyberactivism. They find a trend toward a more integrated use of social media tools and applications such as Twitter, Facebook, and YouTube, generating what could be called "cyberactivism 2.0."[36] What is the situation in China? Examining exemplars will help us better appreciate how ENGOs, new social media, and the environmental movement intersect.

Three Case Studies in ENGO Activism in China

These three cases are by no mean representative of the whole landscape of ENGO activism in the new media, but they provide rich materials to reflect on the current ENGO trends to engage new media and care in their activism. Each one illustrates a different mode of engagement: (1) making culture- and emotion-based appeals; (2)

flooding online sites with public-opinion-based calls to action; and (3) enhancing public deliberation by providing science-focused platforms. Overall, new media is the network used to, as Pezzullo and Cox write, "alert, amplify, and engage" broader public culture.[37] Nevertheless, the particularities of China's context warrant closer examination.

Protecting the Green Peacock as Culture- and Emotion-Based Activism

In March 2017, environmental activist Gu Jianbo discovered the plan to construct a first-grade hydropower station on the Geshai River in Yunnan province.[38] Jianbo spread the news, which quickly caused widespread concern. At the center of attention is a rarely known national level-one endangered species, the green peacock, whose habitat is at the river basin where the dam was to be built. Unlike the commonly known and often seen blue peacock, the green peacock is indigenous to China and is estimated to be more endangered than the giant panda, with less than five hundred still alive.[39] Presumably, concern spread rapidly out of a deep love for the green peacock, which has been an important image in traditional Chinese culture. The green peacock has been imagined as an honored symbol for auspicious and noble quality since ancient times, as seen in the lyrics of an ancient Yuefu folk song: "A pair of peacocks fly southeast, they whirl and linger in each five li" (Kong Que Dong Nan Fei, Wu Li Yi Pai Huai).[40]

Soon after the first message by Jianbo, Xi Zhinong, China's first wildlife photographer and founder of Wild China Film, led his ENGO into the Red River basin for investigation.[41] The photographer and his team captured the precious moments of how wild green peacocks live in the area, providing concrete evidence of potential threats that the proposed hydropower station would have on wild lives. Upon return, Xi Zhinong publicized an article titled "Who Is Killing the Green Peacock? China's Last Intact Habitat for the Green Peacock Is About to Disappear" on his WeChat and Weibo accounts. A campaign to "protect the endangered green peacock" has since been initiated, and soon attracted two other ENGOs to join: FON and the Shanshui (Mountain and Water) Nature Conservation Center.[42] Together, the three ENGOs did research, consulted experts, and jointly issued an urgent proposal to the Ministry of Environmental Protection (MEP), suggesting the halting of the hydropower plan and reassessment of the impact of the plan with a new round of Environmental Impact Assessments (EIA), in order to protect the last home of the green peacock.[43]

For the past two years, the three ENGOs have pushed forward the campaign through judicial procedure, policy advocacy, and environmental education. While the case is still ongoing as we write, some small successes have been achieved: there is increased public attention, the construction plan for the Gejiang first-grade hydropower station is tabled, and the green peacock is listed as an extremely endangered species by the Environmental Protection Bureau of Yunnan Province.[44] As China's most established ENGO, FON played a crucial role in the campaign. In July 2017, FON filed a formal environmental public-interest lawsuit against the China Hydropower Consultant Group Xinping Development Co., Ltd., and the China Electric Construction Group Kunming Survey and Design Institute Co., Ltd., at the local Chuxiong intermediate court of Yunnan province. In August 2017, FON was notified of the successful filing of the case. This event sets the record for China's first preventive public-interest litigation on wildlife protection. On August 28, 2018, the case had its first hearing at Kunming Intermediate People's Court. Due to the complexity of the case, the court will make a decision in the near future, but requested the construction of the hydropower station to be suspended indefinitely.[45]

Besides utilizing its legal expertise, FON has mobilized a strategic online campaign using WeChat as a social media platform. According to the communication expert of FON, this is the first experiment that FON leads a campaign through the WeChat platform. Thus, it is interesting to examine how FON uses WeChat to mobilize the public and advocate for protecting the green peacock—in this case, what kind of features and characteristics does such WeChat communication exhibit? Since April 2017, FON has begun to recruit volunteers on its social media platforms and to organize them into three WeChat groups: the "creative art group," the "science group," and the "activity and communication group."[46] Among the 7,847 chat records collected from three groups, we saw an emerging online discursive space, where volunteers co-create the story of the green peacock in a co-learning and co-discovery process, one we might call fostering an ethic of care.[47]

The green peacock has long existed as symbolically meaningful in Chinese tradition, ecological discourse, and everyday culture. While the uptick in attention emphasizes a scientific understanding of the crisis (i.e., the impact of the dam on the livelihood of the green peacock and other species in the area), mobilization centered around rhetorical appeals of care. Some conversations focused on evidence for the existence of green peacocks in ancient paintings, writings, and other art forms; others discussed representations of green peacocks in daily life, such as clothing. WeChat communication initiated a series of offline activities at times, such as art salons, drawing contests, lectures, and a New Year's party, for which volunteers and the public gathered to exchange creative works that expressed a sense of honor and affection for green peacocks.

Throughout the campaign, FON used WeChat to engage in direct mobilization (36 percent of the posts), calling for volunteers to participate in the campaigns, to gather funding, and for media to attend and participate in spreading news. Such mobilization could have a direct impact on the scale, scope of influence, and concrete campaign results. More interesting to our finding, however, is the remaining indirect mobilization (64 percent of the posts) on cultural and emotional resources. While bearing no direct impact on the success or failure of the campaign, these mobilizations are crucial factors in connecting volunteers culturally and psychologically, thus lubricating, sustaining, and enlarging the campaign.

The cultural resources evoked mainly included popular scientific knowledge related to, and the historical and cultural connotations represented by, the green peacock. In the group chat, the form and color of green peacocks and blue peacocks were constantly compared, and the habitat of green peacocks and their living conditions were noted consistently. Our research finding of a greater emphasis on culture and emotion matches what the communication staff at FON recounted:

> Our approach is to do it through promoting the green peacock as an important cultural symbol of traditional Chinese culture, to put green peacocks on a level of spiritual symbols. From ancient poems and writings, we want to prove that green peacocks exist in traditional culture. Because most people don't know about green peacocks, it's important for us to say that the peacock in traditional culture is actually a green peacock, not a blue peacock. This is very important. Then it is possible that later people can be emotionally touched.[48]

Obtaining the cultural connotations and knowledge of the green peacock gave volunteers a basis for emotional mobilization. A considerable proportion of the group chats showed the volunteers' feelings about the green peacock, be it appreciation, care, love, or worry. A painting of a green peacock may cause volunteers to lament the beauty of the green peacock and worry about its survival; a piece of news on the progress of the judicial process would fill volunteers with exultation and urge them to forward the message to friends. A deep feeling of sympathy for the potential loss of the green peacock abounds, and it is often metaphorically implied as the loss of traditional culture. The WeChat conversation below furnishes such an example:

> A: [picture of green peacock in the wild]
> A: the green peacock is the most beautiful in the wild.
> B: Yeah, yeah.
> A: a large group of them come to the river to drink water. Wow, how beautiful!
> B: the people of Yunnan used to see this scene often?
> A: the environment was good at that time . . . It's not like now [a sad WeChat expression].
> C: Come on, let's make more contributions to our organization.[49]

As shown here, volunteers often post photos of the peacock in the WeChat group, which elicit emotional expressions and comments, and occasionally anger. In this way, they encourage each other, build and maintain emotional ties with each other, and take a more proactive stance to participate in the campaign.

In the WeChat online campaign, FON is inclined to employ emotional and cultural mobilizations. These indirect mobilizations help FON create a virtual space where members engage in discussions, exchange cultural and scientific knowledge, express emotions, and build identity collectively. WeChat communication is strongly oriented toward evoking the green peacock as a (traditional) cultural symbol and the emotional concern of the participants, while combining those with the rational parts of activism, including following legal procedures to act. Such a comprehensive strategy has presented the campaign as a strong one. Framing green peacocks as cultural symbols has reminded the public of the deeper cultural basis upon which environmentalism rests and has connected the volunteers to the endangered peacock on a deeper level. There, volunteers co-construct an indigenous collective identity related to both nature and cultural tradition and sense a deeper mission to protect nature by engaging in other levels of action within civil society.

The Case of Lei Ping as Flooding Online Sites with Public-Opinion-Based Activism

One of the most attention-grabbing environmental figures is Lei Ping, a volunteer for the China Biodiversity Conservation and Green Development Foundation (Lv Fa Hui). She was once reported on by CCTV because of her long-term efforts to protect fireflies. In 2018, she was detained for her environmental advocacy, and a campaign to set her free began. Whereas the green peacock case study focused on cultural and emotional appeals around a threatened animal, our analysis of Lei Ping's story demonstrates the use of public-opinion-based calls to action in online sites.

The story begins in January 2018, when Lei discovered a crisis: a quarry near the National Natural Reserve of Yun Kai Mountain in Xinyi city, Guangdong province, produced serious sewage, dust, and noise pollution. She posted this information online and reported to local news agencies, which led to an investigation by the local government and a rectification notice issued to the quarry and waste incineration plants. On February 23, upon learning that the quarry had been reopened, Lei revisited the scene taking pictures and evidence; she reported again to the local Environmental Protection Bureau and the Forestry Bureau. After a lack of response, on February 28 she posted a report on WeChat and Weibo to reveal the pollution of the quarry in Maoming, hoping to stimulate supervision by public opinion. On March 16, when Lei came back to Xinyi to do a further investigation, she was arrested by the local police. The next day, the Xinyi municipal government announced that "the water body as reported in the online picture [Lei posted] showed no abnormality, thus the content of the picture and texts are seriously inaccurate and caused a bad influence." The local police station then ordered the detention of Lei Ping for "spreading rumor and disrupting public order" for ten days.[50] But, under the unremitting efforts of environmental groups and environmentalists for more than two days, Lei Ping obtained a "Reprieve of Detention" from the Xinyi police station (after being detained for fifty seven hours) and was temporarily released in the afternoon of March 18.[51] Her release was mostly due to the heated communication on WeChat and Weibo during the first few hours, which pressured the local authorities to release the activist.

On March 16, many environmental groups and individuals began to post messages online, expressing shock over her detention. Oftentimes, the messages were deleted by the Internet Supervision department soon after they were out; however, this seemed to foster more posts, causing the widespread attention of netizens and the public. WeChat and Sina Weibo played a major role in pushing forward the progression of the event. We followed the case closely, collected new media messages by screen-grabbing them before they were gone in order to compare and analyze them. We followed and observed ten active WeChat public accounts and eighteen Sina Weibo accounts (run by either an environmental group or individual activists) during March 16 to April 2. We were interested in learning (1) how new media platforms played a role in the progression of this event, and (2) how different new media platforms played different roles in this event.

In our findings, the two social media platforms differ to a large extent in terms of the number, timing, and message content. In terms of the number of the messages, for the ten WeChat accounts, we collected thirty-six articles, which resulted in 148,391 views and 3,024 "likes." The eighteen Weibo accounts resulted in 154,571 fans and 73 original and "retweeted" posts about Lei Ping's case, which were forwarded 1,700 times, and received 575 comments and 575 "likes." Regarding timing, Weibo involved more communication on the first two days of the case, and WeChat gathered more discussion after the release of Lei Ping on March 18, as figure 1 and figure 2 show.

In terms of the content and nature of communication, Weibo reported Lei Ping's case frequently from her detention to release; the main content focused on reporting the event and appealing for her release. Take the Weibo account "Environmental Heavy Case Group No. 6" (Huan Bao Zhong An Liu Zu) as an example. From the beginning, it began to post messages reporting the progression of the event. From March 16 to 18, it published four Weibo posts, reporting the situation and appealing for her release; from March 18 to 20, it continued reporting the case, including her release, and questioned the conditions of her case with four posts. After the 29th, it lost no time covering the comment from the Ministry of Ecology and Environment (MEE) and the filing of Lei Ping's case. Particularly, by 18:58 of March 16, the message about Lei Ping's detention initially posted by this account was forwarded 496 times. As such, Weibo's main communication approach was through frequent, timely, and brief reporting of the event and grabbing public attention. Weibo created the first field of public opinion for Lei Ping's case as well as care for her as a young environmentalist. A post made by "Environmental Heavy Case Group No. 6" on March 18 said:

> This post-90s girl [Lei Ping] was put on trial today. The articles about her detention for protecting the environment were all deleted. I also feel the pressure of fear, but we firmly believe that Lei Ping is innocent. According to the latest news, Lei Ping is still on trial. Friends, please pay attention to the "unjust" case of Lei Ping, she is not guilty.

For WeChat, due to the limit of only one post per day, the total number of posts is much less than that on Weibo. However, each WeChat article is much

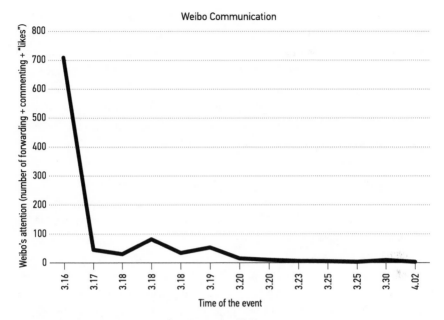

FIGURE 1. Weibo communication of Lei Ping case, 2018. (CREATED BY ZHI WANG)

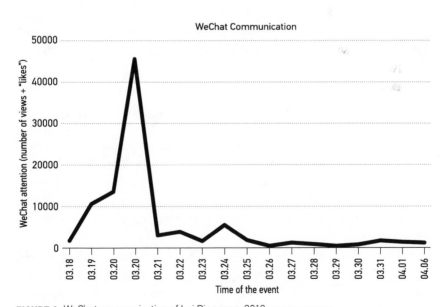

FIGURE 2. WeChat communication of Lei Ping case, 2018. (CREATED BY ZHI WANG)

longer than a Weibo message, and features in-depth discussion and on-the-spot investigation, instead of the brief questioning and emotional venting of Weibo posts. Thus, it takes longer to "ferment the event" on WeChat. The articles that got the most "views" are those posted from March 18 to 20, after Lei Ping was released. Take the WeChat account "Environmental Watch" as an example; it is a platform where all environmental groups and individuals can submit articles to post. Three articles posted there during March 18 to 20 caught more than 60,000 "views"; the article contents focus on in-depth review of and reflection on the event, as well as a field-trip investigation to the disputed quarry. From the 20th to the 29th, "Environmental Watch" continued to post ten articles discussing facts, and urging local authorities to release Lei and reestablish her innocence. One post is based on onsite investigation of the Xinyi Dongjiang River, with numerous photos taken of the polluted scene by another local environmental group confirming that Lei's report was not a rumor but based on hard facts. This particular article was viewed 44,790 times. As such, the WeChat platform became the second new media "fermentation" field, where the public engaged in in-depth discussion and reflection on the event. In one of the WeChat posts, the author quoted President Xi's saying that "we need to protect ecological environment as we protect our eyes," and metaphorically referred to Lei Ping as the eyes through which "we see the exciting beauty of fireflies and painfully mourn for the loss of fireflies," "we see the madness of illegal quarrying and ecological destruction," and "we also see the devotion of grassroots environmental protection and the power of action change." Many discussions and posts like this one appealed for care and love of Lei Ping, the efforts of environmentalists, and the environment in China.

It is worth noting that Weibo and WeChat exhibited different characteristics in addressing the "Lei Ping" controversy due to differences in the platforms. Communication through Weibo tends to be more instantaneous, frequent, spontaneous, rapid, less organized; the messages are succinct, and posting messages is not limited by any gatekeeping mechanism (such as the limit of posting one piece a day). Thus, Weibo communication tends to gather energy quickly and have a strong capacity to stimulate public interest and emotional mobilization. In contrast, WeChat tends to be more organized and thought through, less spontaneous, less frequent, and limited by gatekeeping, which resembles mainstream media production in some ways. WeChat, however, tends to provide the public richer and more in-depth information, creating a space for reflection and rationally based judgments, especially during the fermentation period of the event.

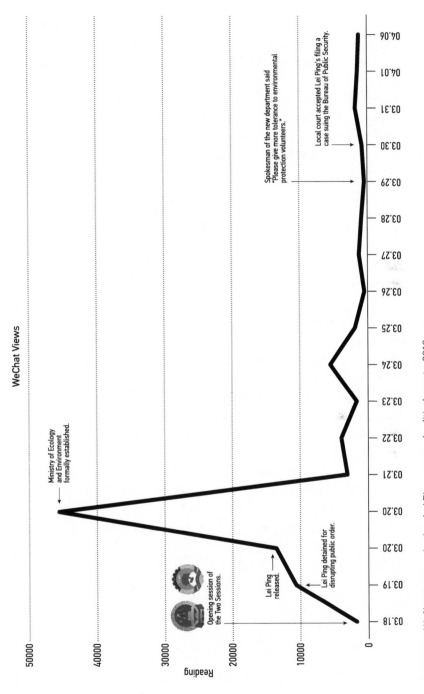

WeChat Views

Reading

50000
40000
30000
20000
10000
0

03.18 03.19 03.20 03.21 03.22 03.23 03.24 03.25 03.26 03.27 03.28 03.29 03.30 03.31 04.01 04.06

Opening session of the Two Sessions.

Lei Ping released.

Lei Ping detained for disrupting public order.

Ministry of Ecology and Environment formally established.

Spokesman of the new department said "Please give more tolerance to environmental protection volunteers."

Local court accepted Lei Ping's filing a case suing the Bureau of Public Security.

FIGURE 3. WeChat communication, Lei Ping case, and political events, 2018. (CREATED BY ZHI WANG)

By sequencing and complementing each other, Weibo and WeChat have been successfully mobilized to assist green groups and environmentalists in achieving their goals. When there was a new development regarding Lei Ping's case, for example, it was Weibo that first reported the news, and then WeChat followed with further discussion and investigation of facts. In this way, ENGOs and activists are able to strategically make use of the different features of these two platforms to maximize the impact of environmental communication and to promote environmental advocacy and advocates in China. They also combine that with a golden political opportunity.

Coincidentally, Lei Ping's case met with two key political events nicknamed "the Two Sessions" that occurred in the same time period, the 13th National People's Congress and the Chinese People's Political Consultative Conference. When Lei Ping posted the pollution message on WeChat on February 28, it was only three days away from the opening session of the Two Sessions. On March 16, when she was detained, the Two Sessions were well underway. On March 19, the MEE was formally established as a new department, and Lei Ping was released the day before. On March 29, Liu Youbin, spokesman of the newly established MEE (indirectly), commented on Lei's case: "We should give more understanding and tolerance to environmental protection organizations and volunteers, giving them more care and support."[52] In something of a U-turn, the next day, the Maonan District People's Court formally accepted Lei Ping's filing of a case suing the Bureau of Public Security, requesting them to cancel her administrative detention.

As figure 3 shows, the pressure of new media, and the grasp of the Two Sessions as a political opportunity are the keys to the release of Lei Ping. What particularly interested us is how different social media platforms played individual and complementary roles during the event. Together, they quickly spread messages that grew exponentially, causing widespread public attention, and formed strong energy and pressure of public opinion in the progression of the event. This synergy helped play a powerful role in publicizing news of Lei Ping and shaping public opinion. Indeed, while traditional news media played nearly no role in publicizing her case, new media networks were used to "alert, amplify, and engage" broader public culture.[53]

IPE's Blue Sky App as Science-Based Activism

The Institute of Public and Environmental Affairs (IPE) and its Blue Sky app offer a case study in science-based activism, for which ENGOs pilot the use of media to spread environmental scientific knowledge and to train citizen scientists. Established by a prominent environmental activist, Ma Jun, in June 2006, IPE is committed to collecting, sorting, and analyzing environmental information released by the government and various enterprises. It built a database of environmental information, a Blue Sky website, and a Blue Sky app. By 2016, IPE's Blue Sky app had been downloaded more than 3 million times.[54] Over the past decade, IPE has made great contributions in promoting public participation in the disclosure of environmental information to protect public health and improve the environment more broadly.

As an ENGO focusing on promoting the disclosure of environmental information, the development of IPE cannot be separated from China's legislation on government information publicity and environmental information disclosure. In 2008, the "Regulation of the Disclosure of Government Information" (RDGI) and the "Measures for the Disclosure of Environmental Information (for Trial Implementation)" (MDEI) were promulgated to make specific provisions on the legal responsibility of governments regarding the disclosure of environmental information and the scope, means, and procedures of making such disclosures. In 2015, the former MEP issued MDEI by Enterprises and Public Institutions, which specified the legal responsibilities, scope, methods, and procedures of the disclosure of environmental information by enterprises and public institutions. The opening of policy space has promoted the continuous development of environmental information disclosure in China, which can also be verified by the increasing amount of environmental information collected and disclosed by the Blue Map (see figure 4).

The Blue Map indicates that year 2008 is a key time point for environmental information disclosure. With the promulgation of RDGI and MDEI, the information collected and released by the Blue Map database in 2008 doubled that of 2007. According to figure 4, between 2008 and 2013, the amount of environmental information disclosed remained at about 20,000 pieces. Since 2014, as the central government further strengthened its requirements on environmental governance, the amount of environmental information disclosed has increased significantly. We also found a total of 324,964 pieces of environmental information released on the

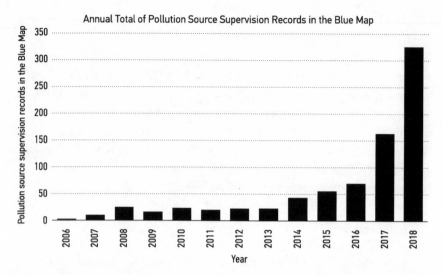

FIGURE 4. Annual pieces of pollution source regulation records released by the Blue Map, 2018.
(CREATED BY IPE, IPE'S ANNUAL REPORT OF POLLUTION INFORMATION TRANSPARENCY INDEX [PITI] FOR 120 CITIES, WWWOA.IPE.ORG.CN/UPLOAD/201904190423050506.
PDF)

Blue Map in 2018, which grew to about 166 times that of 2006. However, although the environmental information collected by the Blue Map shows an impressive growth trend, IPE has been faced with many difficulties during its course of development. As Ma Jun claimed, many local governments have been reluctant to encourage public participation. At one time, environmental information disclosure had been deemed as being very sensitive, which even threatened the survival of IPE.[55] Motivated by local protectionism or economic development, some local governments are not active in disclosing environmental information and may even create obstacles to public participation.[56]

Currently, environmental information released by the Blue Map includes monitoring data of air, water, and soil quality by environmental authorities; environmental supervision information of enterprises; self-monitoring data of enterprises; complaints and reports of the public on environmental issues; and corporate feedback and rectification information. There are two main sources of environmental information for the Blue Map. One is environmental information released by all relevant government departments at all levels, such as departments of water conservancy, land and sea, housing and construction, industry and information technology, development and reform, and meteorology. Another is self-released environmental information by the enterprises. IPE collects information about environmental quality, pollutant emissions, and environmental monitoring from 31 provincial and 338 prefecture-level municipal governments.[57] The database covers 2,402 air-quality monitoring points; 8,022 water-quality monitoring points; and 13,007 polluters.[58] After collecting data from these sources manually or through data-capture software, staff of IPE will store their database and then release it on the Blue Map application and their website. In short, the Blue Map app offers the Chinese public a science-based database, enabling green communication in China to draw upon an immense reservoir of information.

Using this remarkable resource, the public can learn about conventional environmental information, such as air quality and water quality in real time, and they can also look up the environmental performance of enterprises, such as the supervision records of the government and the punishment of enterprises. Meanwhile, the Blue Map makes full use of the interactive features of Web 2.0 and the embedded GIS positioning information, which provides the public a platform to participate in environmental governance more conveniently. The public can take photos of the pollution found in daily life and upload the location and description of the pollution to the Blue Map for real-time reporting. After receiving the report from the public, the staff of IPE will conduct a preliminary examination of the contents of the report and then forward it to the local authorities. According to the Blue Map, the government authorities will reply to the public's report after confirming the situation on the spot. In other words, through new media, the public is not merely a passive recipient of environmental information, but an active participant in monitoring environmental pollution and producing environmental information. The real-time interactive feature of new media opens up the channel for the public to participate in environmental governance. For example, in the Blue Map, there is a section named "protect the good water source in China together with little blue," through which the public can report polluted water sources.

Since 2010, with the promotion of the central and local cyberspace administration departments, party and government agencies have created Weibo and WeChat official accounts. As of June 2015, more than 300,000 certified new media accounts had been created by government agencies at all levels across the country.[59] With the development of new media for government affairs, the public can use new media for commenting, forwarding, and @ing to voice their demands to relevant government departments and report polluting enterprises. The public and ENGOs often use the environmental information released by the Blue Map as evidence to report the pollution on the new media accounts of authorities, so as to promote environmental governance of local governments and the environmental performance of enterprises. Based in Suzhou, Jiangsu province, Green Jiangnan (Lv Se Jiang Nan) is a long-term partner of IPE. As an ENGO focusing on pollution monitoring, Green Jiangnan often refers to environmental information, such as the excessive emissions of enterprises found on the Blue Map, as evidence and then reports it to the relevant environmental protection authorities in the new media platform of governments. This has greatly improved the efficiency of ENGOs participating in environmental governance and promoted the governments' environmental regulation as well as the environmental performance of enterprises.[60]

As an ENGO, IPE's role in promoting environmental information disclosure is more than a collector and "porter" of environmental information. IPE has also actively engaged in pollution monitoring and policy advocacy using the interactive features of Web 2.0. Based on the disclosure performance of five categories of information, including daily supervision of governments, self-monitoring data of enterprises, complaint and reporting of the public, emission data and EIA, IPE annually releases the Pollution Information Transparency Index (PITI), which aims to evaluate the performance in environmental protection of local governments, especially in terms of environmental information disclosure. According to the 2017 report, the PITI covered 120 prefecture-level cities across the country. Ma Jun commented that with the publicity of PITI by IPE and other ENGOs, local governments have come under huge pressure, and an increasing number of them opt to actively cooperate with IPE to improve their performance.[61] In summary, by using new media platforms, IPE successfully expands the channels for the public to participate in environmental governance, promoting the environmental regulation of the government, and the environmental performance of enterprises. While it is science-based, our research reveals how new media-based data collection and circulation can mobilize wider numbers of people to care.

Conclusion

The effect and success of new media-enabled activism is highly context-specific.[62] The Internet (and other ICTs) is embedded within a context affected by the state, the market, civil society, culture, and transnationalism.[63] China may differ from liberal Western democracies in several important ways that limit the potential of new media to function as platforms for critical discourse and positive social change.[64] Nonetheless, new media have exerted a sizable impact, as shown in the three cases addressed herein. They illustrate how Chinese ENGOs use new media to engage the public in cultural and emotional mobilization, to right the misconduct of local governments through producing public opinions, and to spread scientific knowledge. China's activism shows both similarity with its US counterparts and its own environmental advocacy styles due to China's distinct sociopolitical situation.

For example, it is widely observed in the West and the East that emotion-based appeals have strong mobilizing power in contemporary social and political life.[65] Many scholars particularly have found that emotion plays a crucial role in social mobilization and collective action in China.[66] Especially in a (semi-)authoritarian state like China, where rational action lacks sufficient conditions, "emotion is the driving force for Chinese social contention."[67] In extreme cases, China's social movement is even shown as a social anger event and purely an emotional catharsis.[68] To some extent, "saving the green peacock" bears such emotional characteristics. It shows a typical example of how "cultural and emotional" mobilization has been employed in the recent new media activism of Chinese ENGOs. In this campaign, WeChat, as the most popular social media platform in current China, was taken up as a new experimental green space of discourse and mobilization for ENGO activism by fostering an emotionally charged sense of care and drawing upon a long cultural tradition of caring for this species.

In authoritarian political systems, where the organization of collective action tends to be highly restricted, activists also often need to develop extra coping strategies to mobilize public support.[69] Lei Ping's case sets an example of how Chinese ENGOs and activists may strategically utilize various new media platforms at the right time to arouse public attention and to pressure local governments to change decisions on environmental issues. Lei Ping's case may not be representative of new social movements occurring in China, yet such cases do occur once in a while. While communication through new media platforms tends to be unorganized, disordered, and spontaneous, it also has the potential to create energy and pressure. In this case, large numbers of netizens mobilized quickly to flood public opinion about why Chinese people should care about this one individual environmental activist.

As the constitution of environmental problems heavily involves scientific knowledge, science has been a key weapon in the arsenal of environmental activism.[70] It is crucial for social actors to acquire scientific knowledge to successfully make claims on environmental issues. The case of Blue Sky shows the important role of new media in the science-based environmental activism in China. Using new media platforms, IPE greatly improved the efficiency of the public to obtain information and foster public care for nature by appealing to scientific reason and logic. Moreover, advocacy based on the environmental information they collected through new media also promoted the environmental regulation of the government and environmental performance of enterprises.

Chinese online society is bolstered by various social and new media platforms, as almost all ENGOs have set up their own Weibo and WeChat accounts. The recent trend has shown that especially when an environmental accident occurs, new media has become the first space where green groups and activists go to alert public culture about information, amplify events, and mobilize public attention for solutions. Even under restricted sociopolitical conditions, new media continue to empower ENGOs in China, giving opportunities to voice opinions, report events, right governmental misconduct, and push for meaningful social change by mobilizing an ethic of care for environmental communication.

NOTES

The author would like to thank Dongxue Zhong and Zhi Wang for their assistance with collecting data.

1. Abigail R. Jahiel, "The Organization of Environmental Protection in China," *China Quarterly* 156 (1998): 757–787; Peter Ho, "Greening without Conflict? Environmentalism, NGOs and Civil Society in China," *Development and Change* 32 (2001): 893–921; Phillip Stalley and Yang Dongning, "An Emerging Environmental Movement in China?," *China Quarterly* 186 (2006): 333–356; Yanfei Sun and Dingxin Zhao, "Environmental Campaigns," in *Popular Protest in China*, ed. Kevin J. O'Brien (Cambridge, MA: Harvard University Press, 2008), 144–166.

2. Lei Xie, "An Introduction to China," in *Environmental Activism in China* (London: Routledge, 2009), 10–34.

3. According to Regulations on the Registration and Management of Social Associations, and Interim Regulations on Registration Administration of Private Non-enterprise Units, both social associations and private non-enterprise units are forms of social organizations in China. While the former are membership organizations formed voluntarily by citizens, the latter are social organizations established by enterprises, institutions, associations, or other civic entities as well as individual citizens using non-state assets, and conduct nonprofit social service activities. Ministry of Civil Affairs

of the People's Republic of China, "2017 nian shehui fuwu fazhan tongji gongbao" (2017 Statistical Bulletin of Social Service Development), last modified August 2, 2018, http://www.mca.gov.cn/article/sj/tjgb/2017/201708021607.pdf.

4. Guobin Yang, "Environmental NGOs and Institutional Dynamics in China," *China Quarterly* 181 (2005): 46–66.

5. Jingyun Dai and Anthony J. Spires, "Advocacy in an Authoritarian State: How Grassroots Environmental NGOs Influence Local Governments in China," *China Journal* 79, no. 1 (2018): 62–83; Xueyong Zhan and Shui-Yan Tang, "Political Opportunities, Resource Constraints and Policy Advocacy of Environmental NGOs in China," *Public Administration* 91, no. 2 (2013): 381–399.

6. Jianrong Qie, "Huanbao zuzhi huyu huanbaofa zengjia anrijifa deng zhidu" (Environmental organizations called for the introduction of clauses such as punishment on daily basis in Environmental Protection Laws), *Sina.com*, last modified September 28, 2012, http://news.sina.com.cn.

7. Haifang Wang, "Meimei: zhongguo huo tingzhi nujiang daba gongcheng, huanbao renshi ceng kangzheng 10 nian" (US Media: China may shelve the plan of dam construction on Nu River, which the environmentalists have fought for ten years), *Tencent.com*, April 12, 2016, https://new.qq.com/cmsn/20160412/20160412005095.

8. Ho, "Greening without Conflict?," 893–921.

9. Yangzi Sima, "Grassroots Environmental Activism and the Internet: Constructing a Green Public Sphere in China," in *Asian Studies Review* 35, no. 4 (2011): 477–497.

10. Yongkang Wang, "Lvshui qingshan yu Jinshan yinshan" (Lucid waters and lush mountains are golden and silver mountains), *Qiu Shi* (2014): 56–57.

11. Yifu Jia, "'Shishang zuiyan'xin huanbaofa jiangyu 2015 nian 1 yue 1 ri shishi" (The new "strictest" environmental protection law will be implemented on January 1, 2015), *People's Daily Online*, last modified December 28, 2014, http://env.people.com.cn/n/2014/1228/c1010-26287281.html.

12. "Huanjing baohu 'daqi shitiao,' 'shui shitiao,' 'tu shitiao,' ni liaojie duoshao?" (How much do you know about Ten Measures for the Air Pollution, Ten Measures for the Water Pollution, and Ten Measures for the Soil Pollution), *Sina.com*, last modified June 5, 2017, http://nmg.Sina.com.cn/z/graph103/index.shtml.

13. "Guanyu quanmian tuixing hezhang zhi de yijian" (Views on the full implementation of the river chief system), Ministry of Water Resource, last modified December 11, 2016, http://www.mwr.gov.cn/ztpd/gzzt/hzz/zyjs/201708/t20170811_973312.html.

14. Yizhou Fu, "Jiemi zhongyang huanbao ducha zu: chengyuan youshei? Yousha guiju" (Uncovering the Central Environmental Protection Inspection Team), State Council

of China, last modified November 26, 2016, http://www.gov.cn/hudong/2016-11/26/content_5138127.htm.

15. "Top 20 Countries with the Highest Number of Internet Users," Internet World Stats, last modified December 31, 2017, https://www.Internetworldstats.com/top20.htm.

16. "Di 41 ci 'zhongguo hulian wangluo fazhan zhuangkuang tongji baogao" (41st Statistical Report on Internet Development in China), China Internet Network Information Center, last modified March 5, 2018, http://www.cnnic.cn/hlwfzyj/hlwxzbg/hlwtjbg/201803/t20180305_70249.htm.

17. Fengshi Wu and Yang Shen, "Web 2.0 and Political Engagement in China," *VOLUNTAS: International Journal of Voluntary and Nonprofit Organizations* 27, no. 5 (2016): 2055–2076.

18. "Ma Huateng: weixin huoyue yonghushu quanqiu chaoguo 10 yi ren" (Ma Huateng: WeChat has more than 1 billion active users worldwide), *Sina.com*, last modified March 5, 2018, http://tech.Sina.com.cn/roll/2018-03-05/doc-ifyrztfz8050525.shtml.

19. Patrick Shaou-Whea Dodge, "Imagining Dissent: Contesting the Façade of Harmony through Art and the Internet in China," in *Imagining China: Rhetorics of Nationalism in an Age of Globalization*, ed. Stephen J. Hartnett, Lisa B. Keränen, and Donovan Conley (East Lansing: Michigan State University Press, 2017), 311–339.

20. John Arquilla and David Ronfeldt, "The Advent of Netwar (Revisited)," in *Networks and Netwars: The Future of Terror, Crime, and Militancy*, ed. John Arquilla and David Ronfeldt (Santa Monica, CA: Rand, 2001), 1–25; Kjerstin Thorson, Kevin Driscoll, Brian Ekdale, Stephanie Edgerly, Liana Gamber Thompson, Andrew Schrock, Lana Swartz, Emily K. Vraga, and Chris Wells, "YouTube, Twitter and the Occupy Movement: Connecting Content and Circulation Practices," *Information, Communication & Society* 16, no. 3 (2013): 421–451; Phaedra C. Pezzullo and Robert Cox, *Environmental Communication and the Public Sphere*, 5th ed. (Newbury Park, CA: Sage, 2018); Manuel Castells, *Communication Power* (Oxford: Oxford University Press, 2009).

21. Habibul Haque Khondker, "Role of the New Media in the Arab Spring," *Globalizations* 8, no. 5 (2011): 675–679; Merlyna Lim, "Clicks, Cabs, and Coffee Houses: Social Media and Oppositional Movements in Egypt, 2004–2011," *Journal of Communication* 62, no. 2 (2012): 231–248; Richard Kahn and Douglas Kellner, "New Media and Internet Activism: From the 'Battle of Seattle' to Blogging," *New Media & Society* 6, no. 1 (2004): 87–95; Jenny Pickerill, *Cyberprotest: Environmental Activism Online* (Manchester: Manchester University Press, 2013); Castells, *Communication Power*, 2009; Luis E. Hestres, "Preaching to the Choir: Internet-Mediated Advocacy, Issue Public Mobilization, and Climate Change," *New Media & Society* 16, no. 2 (2014): 323–339; Summer Harlow and Dustin

Harp, "Collective Action on the Web: A Cross-Cultural Study of Social Networking Sites and Online and Offline Activism in the United States and Latin America," *Information, Communication & Society* 15, no. 2 (2012): 196–216; Wojcieszak Magdalena, "'Carrying Online Participation Offline'—Mobilization by Radical Online Groups and Politically Dissimilar Offline Ties," *Journal of Communication* 59, no. 3 (2009): 564–586.

22. Bruce Bimber, Andrew J. Flanagin, and Cynthia Stohl, "Reconceptualizing Collective Action in the Contemporary Media Environment," *Communication Theory* 15, no. 4 (2005): 365–388; Lance W. Bennett, "New Media Power: The Internet and Global Activism," in *Contesting Media Power: Alternative Media in a Networked World*, ed. Nick Couldry and James Curran (Lanham, MD: Rowman & Littlefield Publishers, 2003), 17–37.

23. David Ronfeldt and John Arquilla, "Emergence and Influence of the Zapatista Social Netwar," in Arquilla and Ronfeldt, eds., *Networks and Netwars*, 171–199; Castells, *Communication Power*, 333.

24. Arquilla and Ronfeldt, "The Advent of Netwar (Revisited)," 1–25; Dieter Rucht, "The Quadruple 'A': Media Strategies of Protest Movements since the 1960s," in *Cyberprotest: New Media, Citizens, and Social Movements*, ed. Wim van de Donk, Brian D. Loader, Paul G. Nixon, Dieter Rucht (New York: Routledge, 2004), 29–56; W. Lance Bennett and Alexandra Segerberg, "The Logic of Connective Action: Digital Media and the Personalization of Contentious Politics," *Information, Communication & Society* 15, no. 5 (2012): 739–768; Peter van Aelst and Stefaan Walgrave, "New Media, New Movements? The Role of the Internet in Shaping the 'Anti-globalization' Movement," *Information, Communication & Society* 5, no. 4 (2002): 465–493; Julie Uldam, "Activism and the Online Mediation Opportunity Structure: Attempts to Impact Global Climate Change Policies?," *Policy & Internet* 5, no. 1 (2013): 56–75; Steve Wright, "Informing, Communicating and ICTs in Contemporary Anti-Capitalist Movements," in *Cyberprotest: New Media, Citizens, and Social Movements*, ed. Wim van de Donk, Brian D. Loader, Paul G. Nixon, and Dieter Rucht (New York: Routledge, 2004), 77–93; Joanne Lebert, "Wiring Human Rights Activism: Amnesty International and the Challenges of Information and Communication Technologies," in *Cyberactivism: Online Activism in Theory and Practice*, ed. Martha McCaughey and Michael D. Ayers (New York: Routledge, 2003), 209–231; Bennett, "New Media Power," 17–37.

25. Khondker, "Role of the New Media," 675–679.

26. Lim, "Clicks, Cabs, and Coffee Houses," 231–248.

27. Yongnian Zheng and Guoguang Wu, "Information Technology, Public Space, and Collective Action in China," *Comparative Political Studies* 38, no. 5 (2005): 507–536.

28. Pezzullo and Cox, *Environmental Communication and the Public Sphere*, 2018; Wenhong

Chen and Stephen D. Reese, eds., *Networked China: Global Dynamics of Digital Media and Civic Engagement: New Agendas in Communication* (New York: Routledge, 2015); Jonathan Sullivan and Lei Xie, "Environmental Activism, Social Networks and the Internet," *China Quarterly* 198 (2009): 422–432, passim.

29. Jonathan Sullivan, "A Tale of Two Microblogs in China," *Media, Culture & Society* 34, no. 6 (2012): 773–783.

30. Jingfang Liu, "Picturing a Green Virtual Public Space for Social Change: A Study of Internet Activism and Web-based Environmental Collective Actions in China," *Chinese Journal of Communication* 4, no. 2 (2011): 137–166; Guobing Yang, "Contention in Cyberspace," in *Popular Protest in China*, ed. Kevin J. O'Brien (Cambridge, MA: Harvard University Press, 2008), 126–143.

31. Yang, "Contention in Cyberspace," 126.

32. Liu, "Picturing a Green Virtual Public Space," 137–166.

33. Sima, "Grassroots Environmental Activism," 477–497.

34. David W. Wilson, Xiaolin Lin, Phil Longstreet, and Saonee Sarker, "Web 2.0: A Definition, Literature Review, and Directions for Future Research," in *AMCIS 2011 Proceedings—All Submissions*, Paper 368 (2011), http://aisel.aisnet.org/amcis2011_submissions/368.

35. Wu and Shen, "Web 2.0," 1–22.

36. Rodrigo Sandoval-Almazan and J. Ramon Gil-Garcia, "Towards Cyberactivism 2.0? Understanding the Use of Social Media and Other Information Technologies for Political Activism and Social Movements," *Government Information Quarterly* 31, no. 3 (2014): 365–378.

37. Pezzullo and Cox, *Environmental Communication*, 232.

38. WildChina, "Shi shui zai 'shasi'lvkongque? Zhongguo zuihou yipian lvkongque wanzheng qixidi jijiang xiaoshi" (Who is "killing" the green peacock? The last intact habitat of green peacock is disappearing), *blog.sina.com*, last modified November 8, 2017, http://blog.sina.com.cn/s/blog_740df2450102xl94.html.

39. Blue peacock, renamed India peacock, originated in India and Sri Lanka and is the national bird of India. Green peacock, class I national protection wild bird, is listed as a globally endangered species by the International Union for Conservation of Nature and under Appendix 2 of "Convention on International Trade in Endangered Species." Minghui Xu, "Lvkongque zhongqun shuliang buzu 500 zhi, qing baohu hao tamen de qixidi" (The population size of green peacock is estimated to be less than 500; Please protect their habitat), *The Paper.com*, last modified December 3, 2017, http://www.thepaper.cn/newsDetail_forward_1653548.

40. The "li" is a traditional Chinese measurement for distance, it is smaller than the English

mile.

41. A renowned NGO in China dedicated to protecting and recording China's disappearing nature and wildlife through photography; http://www.wildchina.cn.

42. ShanShui Nature Conservation Center was initiated in 2007 as a grassroots ENGO, and is dedicated to nature conservation practices of native Chinese society and culture; http://www.shanshui.org.

43. China Wildlife Conservation Association, "Wode 2017 Lvkongque baoweizhan 2018 qing he women yiqi xingdong" (My 2017 Green Peacock Battle, join us to act in 2018), *Sohu.com*, last modified January 18, 2018, https://www.sohu.com/a/217413735_261762.

44. Qiaochu Huang, "Baohu lvkongque yufangxing gongyi susongan kaiting, shuidianzhan jianshefang: yi wufa kaigong" (Preventive public interest litigation for protection of the green peacock has opened a court session, Hydropower Station Construction Unit: It could not be started), *The Paper.com*, last modified August 29,2018, https://www.thepaper.cn/newsDetail_forward_2389689.

45. Ibid.

46. By the time we finished observing the groups on April 30, 2018, there were 55, 56, and 64 members in each group respectively. The three groups are not exclusive; members from one group may join other groups if they like.

47. In an online ethnography research of these WeChat groups, we joined the groups, and observed and studied the WeChat communication from August 2017 to April 2018 for nine months. Among 7,847 chat records, 2,521 were made by the creative art group, 2,254 were from the science group, and the remaining 3,072 were from activity and communication group.

48. Interview conducted in August 2017.

49. WeChat "friend circle" communication, November 13, 2017.

50. Xinyi Municipal Government, "Xinyishi yifa chachu yiqi sanbu yaoyan raoluan gonggong zhixu anjian" (Xinyi City investigated and punished a case of spreading rumors and disturbing public order)," *Xinyi.gov.cn*, last modified March 17, 2018, http://www.xinyi.gov.cn/ywdt/bmdt/content_15474.

51. Nanxi Zhang, "Huanbao renshi baoguang wuran beiju xu: zanhuan juliu dangshiren chen buyanjin dan weizaoyao" (Environmentalist detention: Suspension of detention, imprecise but not rumored), *sohu.com*, last modified March 23, 2018, http://www.sohu.com/a/226179360_658437.

52. Yingchun He, "Shengtai huanjing bu huiying zhiyuanzhe beiju: ying duo yixie lijie he baorong" (Ministry of Ecology and Environment responded to volunteer detention: More understanding and tolerance), *env.people.com*, last modified March 29, 2018, http://env.

people.com.cn/n1/2018/0329/c1010-29896845.html.

53. Mainstream media fell short of its role in this case. Only four news articles with cautious reporting from three traditional media (the *Phenix News*, *The Paper*, and the China Jiangxi Net) were found. The brief reporting only covered the detention of Lei and the comment of the spokesperson from MEE, without following up on her release. Pezzullo and Cox, *Environmental Communication*, 232.

54. IPE's Annual Report of 2016, http://wwwoa.ipe.org.cn//Upload/201703101152121174.pdf.

55. Interview conducted on October 24, 2017.

56. Director of ENGO Green Jiangnan, interview conducted on July 15, 2018.

57. "About Us," IPE, http://www.ipe.org.cn/about/about.aspx.

58. IPE's Annual Report of 2016.

59. "2015 nian 1–6 yue quanguo zhengwu xinmeiti zonghe yingxiangli baogao" (Report on the comprehensive influence of government new media in China from January to June 2015 was released), *Xinhua Net*, last modified August 18, 2015, http://www.xinhuanet.com/yuqing/2015-08/18/c_128137211.htm.

60. Director of Green Jiangnan, interview conducted on July 15, 2018.

61. Interview conducted on October 24, 2017.

62. Ronggui Huang and Ngai-ming Yip, "Internet and Activism in Urban China: A Case Study of Protests in Xiamen and Panyu," *Journal of Comparative Asian Development* 11, no. 2 (2012): 201–223.

63. Guobin Yang, *The Power of the Internet in China: Citizen Activism Online* (New York: Columbia University Press, 2009).

64. Gary King, Jennifer Pan, and Margaret E. Roberts, "How Censorship in China Allows Government Criticism but Silences Collective Expression," *American Political Science Review* 107, no. 2 (2013): 326–343; Rebecca MacKinnon, "China's 'Networked Authoritarianism,'" *Journal of Democracy* 22, no. 2 (2011): 32–46; Josh Chin, "China Is Requiring People to Register Real Names for Some Internet Services," *Wall Street Journal*, February 4, 2015.

65. Omar V. Rosas and Javier Serrano-Puche, "New Media and the Emotional Public Sphere," *International Journal of Communication*, no. 12 (2018): 2031–2039; Yu Hong, "Internet Mobilization: How to Achieve from Virtual to Reality?," *Southeast Communication*, no. 1 (2010): 74–75; Zuying Xu, "Internet Mobilization Analysis Review—From Conflict Management Perspective," *Journal of Yanshan University* (Philosophy and Social Science Edition), no. 2 (2015): 72–78.

66. Jinhong Wang and Zhenhui Huang, "Zhongguo ruoshi qunti de beiqing kangzheng jiqi lilun jieshi—yi nongmin jiti xiagui shijian wei zhongdian de shizhengfenxi" (The

pessimistic battle and theoretical explanation of Chinese vulnerable group—Take farmers collectively kneeling down as an important empirical analysis), in *Journal of Sun Yat-sen University* (Social Science Edition), no. 52 (2012): 152–164; Guangfeng Yuan, "Gonggong yulun jiangou zhong de 'ruoshi gan'—jiyu 'qinggan jiegou' de fenxi," *Journalism Review*, no. 4 (2015): 47–53.

67. Xiao'an Guo, "The Choice and Mobilization Effect between Sanity and Emotion in Social Conflicts—Based on Statistical Analysis of 120 Events in 10 Years," *Chinese Journal of Journalism and Communication*, no. 39 (2017): 107–125.

68. Jianrong Yu, "The Current Major Category and Basic Characteristics in China Group Events," *Journal of China University of Political Science and Law*, no. 6 (2009): 114–120; Xing Ying, "'Aura' and Generative Mechanism of Group Events—Comparison between Two Individual Cases," *Sociology Study*, no. 6 (2009): 105–121; Wei Sun, "'Who Are We': The Collective Identity Construct on New Social Movements by Mass Media—Case Analysis on Mass Media Reports of Xiamen PX Project Event," *Journalism Quarterly*, no. 3 (2007): 140–148.

69. Andrew Wells-Dang, *Civil Society Networks in China and Vietnam: Informal Pathbreakers in Health and the Environment* (New York: Palgrave Macmillan, 2012).

70. Charlotte Epstein, "Knowledge and Power in Global Environmental Activism," *International Journal of Peace Studies*, no. 10 (2005): 47–67.

PART TWO

On Crisis

Comparing Chinese and American Public Opinion about Climate Change

Binbin Wang and Qinnan (Sharon) Zhou

C limate change is one of the most—if not *the* most—pressing crises of our times. In a historic attempt to assess public opinion about this crisis, the China Center for Climate Change Communication (China4C) and the Yale Program on Climate Change Communication (YPCCC) cohosted an official event titled "Comparing and Contrasting Public Opinion about Climate Change in China and the United States" during the 2017 United Nations Climate Change Conference (COP23). As part of our efforts, the China4C and YPCCC jointly released two reports, *Climate Change in the American Mind* and *Climate Change in the Chinese Mind*, which summarized the then-latest public surveys gauging people's awareness of and beliefs about climate change in the world's two largest CO_2 emitters.[1] In this chapter, we summarize the findings conveyed in these reports and provide some observations on what they suggest for the future of US-China relations, especially regarding the questions of climate change, citizen awareness, and public willingness to respond.

The event name suggests connections between the two surveys: some questions, particularly those covering beliefs and personal experiences, and how media engagements from one nation mirror those in the other; other questions diverge to examine how the country-specific demographic characteristics and social scenarios affect people's beliefs. Some survey results are similarly positive in both countries, including the finding that most surveyed Americans and Chinese believe climate change or global warming is happening, and that large percentages of respondents in both countries indicate that they're increasingly worried about it. Our data, however, also suggests that the emotional responses and actual willingness to act on climate change vary between Chinese and American respondents.

One wording difference worth pointing out before taking a closer look at the two surveys' results is the keyword choice of "climate change" and/or "global warming." Following the framing established in their signature 2008 joint study, *Global Warming's Six Americas*, the YPCCC's and the George Mason Center for Climate Change Communication's (4C) survey consistently uses "global warming." In comparison, China4C's two national surveys in 2012 and 2017 both use the term "climate change." As shown in the YPCCC and 4C's 2014 report "What's in a Name?," "global warming" and "climate change" are often not synonymous: "they mean different things to different people and activate different sets of beliefs, feelings, and behaviors, as well as different degrees of urgency about the need to respond."[2] As they clarify, "climate change" appeared in scientific literature as early as 1956, whereas "global warming" did not appear until 1975. Although often used interchangeably in mainstream media, working definitions do reveal a distinction:

> *Global warming* refers to the increase in the Earth's average surface temperature since the Industrial Revolution, primarily due to the emission of greenhouse gases from the burning of fossil fuels and land use change, whereas *climate change* refers to the long-term change of the Earth's climate including changes in temperature, precipitation, and wind patterns over a period of several decades or longer.[3]

Interestingly, Americans have historically used "global warming" as a Google Internet search term more frequently than "climate change."[4] Further, global mass media widely adopted "global warming" beginning in the late 1980s, when the term gained popularity in the United States; in contrast, most policies and action plans issued by the Chinese government address "climate change." Thus, the different choice of terms is likely to generate greater resonance among the publics addressed by each survey and, therefore, reveal more representative survey results within and across these two cultures.

The scope of these two surveys is also different. YPCCC and 4C are strictly interested in asking questions pertaining to how *global warming* is perceived among Americans, both rationally and emotionally, as well as what might possibly cause these reactions. China4C, however, aims to cover broader attitudes through questions about, but not limited to, *climate change*, including Chinese attitudes toward lifestyle changes, government policies, and international cooperation on this issue. This methodological difference is important to note, as the structures of the two surveys led to different types of questions, with the American-led surveys narrowly focused on "global warming" and the Chinese-led surveys including more broadly construed topics, which aimed to cultivate a more diverse data set.

In this chapter, we offer summaries of these two reports, with our comments focused along three comparison points, which offer a cross-cultural series of snapshots about Chinese and American public opinions about environmental communication. These comparisons focus on the 2017 reports, as well as more recent reactions to the Paris Climate Agreement and US President Donald Trump's declaration of withdrawal from that landmark accord. Overall, we hope this contribution promotes insight and further discussion about the similarities and differences in public opinion about the climate-change crisis among people living in China and the United States.

Climate Change Beliefs

In this section of the chapter, we offer three points of comparison to consider public opinion about the existence of global warming or climate change. First, we consider public opinions about whether or not global warming/climate change exists. Second, we offer an assessment of the levels of worry. Third, we summarize public opinion about whether or not climate change/global warming is caused by human activity.

In response to a series of questions exploring people's beliefs, basic knowledge, and concerns over climate change/global warming, we asked if survey participants believed climate change/global warming existed. As you can see on the left side of figure 1, a higher percentage of Chinese (94.4 percent) we surveyed believe that it is happening than Americans (71 percent). Another way to state this finding is that more than 9 in 10 Chinese think global warming/climate change is happening, while only 7 in 10 Americans agree. In both cases, the majority believe the climate science consensus that global warming/climate change has begun, but the majority is not as large in the United States. Further, while the percentage

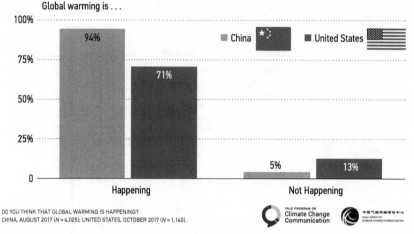

FIGURE 1. Belief in existence of climate change, China and United States comparison

of "deniers" accounts for only 5.3 percent of Chinese respondents, 13 percent of American respondents could be categorized as denying that climate change/ global warming exists. Disbelief in the climate science consensus, therefore, is statistically more of a barrier for climate action and policy in the United States than China—though the clear majority in both countries think climate change/ global warming is happening.

Moving from figure 1 to figure 2, our second major finding concerns whether or not there is worry about climate change/global warming in China and the United States. Almost four in five Chinese respondents and two-thirds of surveyed Americans are at least "somewhat worried about climate change/global warming." These results correlate with our previous finding that most people in both countries are concerned about climate change as a crisis we face globally. Notably, however, anxiety is more intense among Americans, as one in five (22 percent) are "very worried" in comparison with 16.3 percent Chinese. This is an interesting figure, for while the percentage of Americans acknowledging the reality of climate change was lower than in China, the percentage of Americans reporting their being "very worried" was higher than in China, indicating a wide swing in political beliefs. Indeed, while more Americans deny the problem than in China, more Americans are "very worried" than in China—the range of political opinions about climate

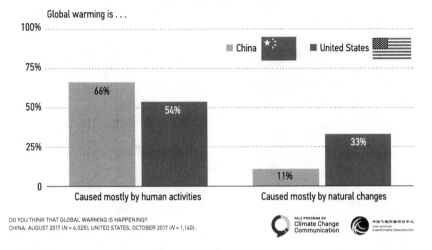

Two in Three Chinese Think Global Warming Is Mostly Human Caused
— More than Half of Americans Think So —

FIGURE 2. Worries about climate change, China and United States comparison

change is therefore more diverse, more widely fractured in America than in China, reflecting the polarized nature of political discourse in the United States.

American respondents also were queried about their emotions. The majority feel they are "interested" (67 percent), and more than half feel "disgusted" (55 percent) or "helpless" (52 percent). This typology of concern has become fine-tuned as a key finding in the United States in the 2018 report *Engaging Diverse Audiences with Climate Change: Message Strategies for Global Warming's Six Americas.* There, our US colleagues at YPCCC and 4C identify the following audiences for climate-change communication: Alarmed (21 percent); Concerned (30 percent); Cautious (21 percent); Disengaged (7 percent); Doubtful (12 percent); and Dismissive (9 percent).[5] Given the larger percentage of denial in the United States, this more specific typology reinforces YPCCC's finding that the majority of Americans (72 percent) are alarmed, concerned, or cautious. The fact that 28 percent of American respondents are disengaged, doubtful, or dismissive again reflects, we assume, larger political patterns in a nation where engagement with the political process is uneven.

An interesting observation related to these findings is our third data point about whether or not climate change/global warming is anthropogenic or human-caused. Over half (64.3 percent) of Chinese respondents say they know "just a little about it" or "have never heard of it"; yet, 66 percent of Chinese (versus 54 percent

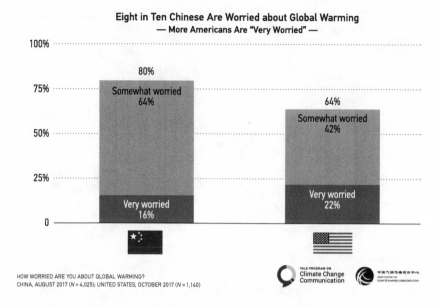

FIGURE 3. Anthropogenic or human-caused climate change acceptance, China and United States comparison

of American) respondents understand that climate change/global warming is mostly "human-caused" (see figure 3). Many Chinese respondents also correctly associated climate change with some of the negative climate phenomena science has documented, such as "hot," "haze," and "global warming."

Now that we have compared and contrasted basic climate-change/global-warming public opinion, we want to turn more to the perceptions of these impacts before returning to the topic of communication more explicitly.

Perceptions of Climate Change Impacts

Most Chinese and Americans tend to view climate change/global warming as a more distant threat (see figure 4). While one-third of Chinese and nearly half of American respondents think climate change/global warming will harm themselves and/or their families at least to a moderate degree, the percentages increase to 78 percent for Chinese and 75 percent for American respondents, respectively, when being asked about harm to future generations.

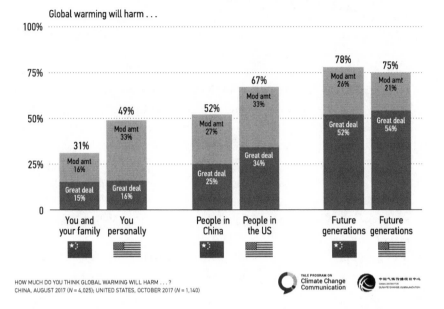

FIGURE 4. Perception of global warming harms, China and United States comparison

Notably, in October 2017, almost half of Chinese (52 percent) perceived their country as being harmed "right now" by global warming/climate change, while 67 percent of Americans felt people in the United States were being harmed "right now." The number of Americans that felt a more immediate effect on their families, therefore, correlated closer (within 18 percent) with whether or not they perceived that the United States in general was feeling the impact. For Chinese people, however, relatively more (36 percent) believed someone in China was being impacted even when their own family didn't feel the impact already. Thus, we at China4C found that the Chinese public tends to consider climate change a more distant problem and is less likely to associate negative impacts of climate change with what is happening presently.

Further, in the YPCCC and 4C's 2017 survey, several extra questions were included to assess people's in-depth understanding of the relationships between climate change and extreme weather events. For example, nearly two in three Americans think global warming is affecting weather in the United States, and a majority of them think global warming exacerbated several extreme events in

2017, such as the heat waves in California (55 percent) and Arizona (51 percent); Hurricanes Harvey, Irma, and Maria (54 percent); and wildfires in the western United States (52 percent). Considering future impacts, many Americans have recognized that global warming will cause melting glaciers (67 percent) and worsen potential disasters, including severe heat waves (64 percent), droughts and water shortages (63 percent), and floods (61 percent) over the next two decades. These data points offer a curious complication in our analysis, for while Americans seem to understand that global warming is causing extreme weather and weather-related disasters now, their chief concerns about global warming, as noted above, appear to target the future. As is true in large surveys about health, national security, and economics, then, our data reveals a fundamental irrationality in how respondents piece together their daily experiences (the weather is going crazy now) and their overarching belief systems (but global warming will cause more damage later).

Likewise, the Chinese public identify several impacts of climate change over the next twenty years in China: disease epidemics (91.3 percent), droughts and water shortages (89.8 percent), floods (88.2 percent), glaciers melting (88.0 percent), extinctions of plant and animal species (83.4 percent), and famines and food shortages (73.4 percent). Chinese people almost unanimously (95.1 percent) think climate change will cause air pollution, the biggest concern (33.4 percent) among all the above unfavorable impacts. In follow-up questions asking which fields the central government should pay special attention to, the Chinese public again believe the government should give air pollution the highest priority, followed by ecological protection (18.0 percent) and health issues (17.2 percent). As these numbers indicate, high percentages of Chinese respondents understand that global warming is directly linked to impacts on both daily experiences (air, water, food, and so on) and extreme events (floods, droughts, and species extinction).

Responding to and Communicating about Climate Change

The YPCCC and 4C 2017 survey show that most Americans think global warming is an environmental/scientific issue, and nearly half of Americans say humans could reduce current climate-change trends; however, general opinion tends to believe it's unclear "at this point whether we will do what is necessary" to respond to global warming/climate change.[6] The survey didn't lay out any specific action options to reduce emissions, nor did it ask about the American public's attitude and willingness to act. Most surveyed Americans (78 percent) expressed their interest in learning more about global warming, among which 22 percent are "very" interested, which may indicate the potential for them to adopt a less carbon-intensive lifestyle once they know more about the issue and are provided with practical options for action. This indicates an area that needs future research.

In comparison, the China4C survey scrutinizes the Chinese public's attitudes toward both climate policies and climate actions in great detail. In keeping with the tenor of politics in China, most Chinese respondents think that the government should play the leading role and shoulder more responsibilities in combating climate change. This belief may explain the generally high support level (above 90 percent) for the government's efforts at controlling the country's emissions and both mitigating and adapting to climate change. When it comes to personal actions, 73.7 percent of respondents are willing to pay more for climate-friendly products, and nearly one-third are willing to pay RMB 200 (Chinese yuan) annually to offset their personal emissions. Bike sharing, as a trendy, low-carbon lifestyle, is enjoying widespread support in China (92.6 percent) and nearly half of the respondents have already used shared bikes. Choices about climate action appear welcome and to be emerging both personally and for the national leadership. These findings confirm a can-do attitude, a sense that combining personal actions with governmental leadership can lead to positive outcomes.

Communication is essential to advocating climate policy with the government or to promoting individual lifestyle choices. In both China and the United States, we find that media provide the largest source of climate-change/global-warming information. In America, over half of the respondents surveyed had heard about global warming in the media at least once a month. The US survey shows that 38 percent of Americans discuss global warming with family and friends at least occasionally, and one in three Americans share relevant news stories about global warming on social media. In China, television (83.6 percent), WeChat (79.4 percent), and friends and family (68.1 percent) are the top three news sources reported. The China4C survey doesn't specifically ask if and how often the Chinese public proactively share climate change information in real life, but the results reflect that Chinese people (97.7 percent) are more than willing to circulate stories. The Chinese report also shows a much stronger reliance on social media (mainly WeChat) for Chinese people compared to their American counterparts in acquiring climate change information. Additionally, almost all (94 percent) Chinese have a strong desire to learn more about climate change, and they (98.7 percent) demand schools teach students about its causes, consequences, and solutions. This is a striking finding for future educational policy and the growth of climate change communication as a field in China.

International Cooperation on Climate Change

Now that we have established public opinion about climate-change/global-warming causes, impacts, responses, and communication, we want to end by thinking about the Trump administration's impact on public opinion in both countries. Taking office in January 2017, in part with a strong anti-climate-policy platform, President Trump has had a detrimental impact on public opinion globally. Conversely, Chinese President Xi Jinping has repeatedly pledged that China will continue to take steps to tackle climate change and to fulfill its obligations under the United Nations climate accord. Particularly at the 19th Congress of the Chinese Communist Party in October 2017, Xi said that China has been "an important participant, contributor, and torchbearer in the global endeavor for ecological civilization" by taking the lead in the "international cooperation" on climate change. China's commitment to combating climate change offers hope in the continued advancement of international climate-governance agendas, despite recent reversals in public opinion precipitated by America's forfeiture of its former leadership position. In both countries, of course, we need to be cautious and constantly scrutinize whether the actual climate policies and practices align with presidential decrees.

Still, as we opened this chapter with the significant 2015 COP21 gathering, it is important to revisit this historic milestone in light of recent events. The United States' decision in 2017 to withdraw from the Paris Agreement has caused uncertainties in global climate governance.[7] However, Chinese high support (96.3 percent) of the country's participation in the Paris Agreement remains unwavering, even as the US government tries to roll back climate policies (see figure 5). Indeed, a high majority (94 percent) of respondents think China should stay in the Paris Agreement for the purpose of curbing climate-induced pollution, and 96.8 percent of them approve of China's effort to promote international climate collaborations through, for example, initiatives assisting poorer developing countries under the framework of the South-South Cooperation on Climate

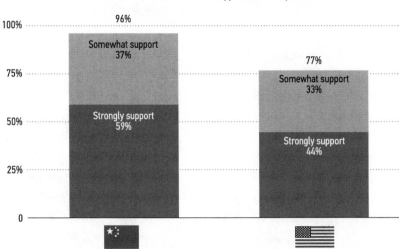

FIGURE 5. Support for participation in Paris Climate Agreement, China and United States comparison

Change. This finding indicates that the current US government's climate policy standpoint has had little impact on China's global leadership and perhaps has strengthened the nation's resolve.

Though the original US survey report didn't directly probe into whether Americans are supportive of the current administration's decision, a 2017 YPCCC and 4C report built on the nationally representative survey, demonstrating that the majority of American voters disagree with Trump's retreat from global governance. Indeed, 64 percent of registered voters at least somewhat oppose and 45 percent strongly oppose the decision to pull out of the Paris Accord.[8] Another inspiring result is that a strong majority, three in four registered American voters (77 percent), think that the United States should participate in the Paris Climate Agreement. These data points indicate that President Trump is making decisions about climate issues that do not reflect American public opinion.

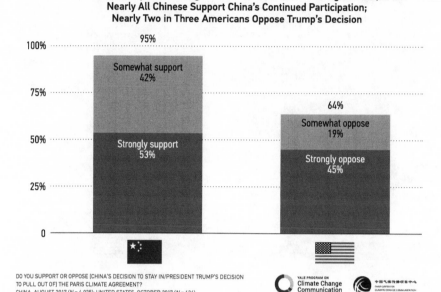

With Trump Announcing He Will Pull Out of the Paris Agreement,
Nearly All Chinese Support China's Continued Participation;
Nearly Two in Three Americans Oppose Trump's Decision

DO YOU SUPPORT OR OPPOSE [CHINA'S DECISION TO STAY IN/PRESIDENT TRUMP'S DECISION TO PULL OUT OF] THE PARIS CLIMATE AGREEMENT?
CHINA, AUGUST 2017 (N = 4,025); UNITED STATES, OCTOBER 2017 (N = 626)

FIGURE 6. Response to Trump's withdrawal from Paris Climate Agreement, China and United States comparison

Conclusion

These two surveys demonstrate that the majority of the public in China and the United States support climate actions and the transition to a more sustainable future. These encouraging insights should help inform policymakers and civil society in both countries to reaffirm their climate commitments and to mobilize the public to act in response to this crisis. We have indicated that both the United States and China have more work to do with climate communication, particularly in the areas of in-depth environmental education and identifying feasible climate actions. We also found that the vast majority of Chinese people want greater climate communication education, as well as action. Across the board, we believe our data should be inspiring for climate educators and activists, for the majority of citizens in both countries want to work—and to work together—on building a safer, healthier, greener future. From this perspective, we might conclude that green communication can and should stand at the heart of future US-China relations.

NOTES

1. Anthony Leiserowitz et al., "Climate Change in the American Mind: October 2017" (New Haven, CT: Yale Program on Climate Change Communication: Yale University and George Mason University, 2017); B. Wang et al., "Climate Change in the Chinese Mind: Survey Report 2017" (China Center for Climate Change Communication, 2017). On the latter, see China Center for Climate Change Communication, "China4C's 2017 National Public Opinion Survey Report Climate Change in the Chinese Mind Released at COP23," UNFCCC COP23 press release (2017a), https://unfccc.int/sites/default/files/resource/Press%20Release%20-%202.pdf.

2. Anthony Leiserowitz et al., "What's in a Name? Global Warming vs. Climate Change" (New Haven, CT: Yale Project on Climate Change Communication, Yale University and George Mason University, 2014), http://environment.yale.edu/climate-communication-OFF/files/Global_Warming_vs_Climate_Change_Report.pdf, 4.

3. Ibid., 6.

4. Ibid., 8.

5. "Global Warming's Six Americas," Yale Program on Climate Change Communication, November 1, 2016, http://climatecommunication.yale.edu/about/projects/global-warmings-six-americas.

6. Connie Roser-Renouf et al., "Engaging Diverse Audiences with Climate Change: Message Strategies for Global Warming's Six Americas" (Fairfax, VA: George Mason University Center for Climate Change Communication, 2014), https://climatecommunication.yale.edu/publications/global-warmings-six-americas-book-chapter-preview; for another version of this report, see Anders Hansen and Robert Cox, eds., *Routledge Handbook of Environment and Communication* (London: Routledge, 2015).

7. China Center for Climate Change Communication, "A Majority of the Public in China and the US Support the Paris Agreement and the Transition to Clean Energy, New Survey Findings Show," UNFCCC COP23 press release (2017b), https://climatecommunication.yale.edu/news-events/press-release-majority-public-china-us-support-paris-agreement-transition-clean-energy-new-survey-findings-show.

8. Anthony Leiserowitz et al., "By More than 5 to 1, Voters Say the US Should Participate in the Paris Climate Agreement" (New Haven, CT: Yale Program on Climate Change Communication: Yale University and George Mason University, 2017); for report, also see China Center for Climate Change Communication, "New Surveys Show a Majority of Americans and Chinese Support the Paris Agreement and the Transition to a Low Carbon Future," UNFCCC COP23 press release (2017c), https://unfccc.int/sites/default/files/resource/Press%20Release.pdf.

Examining Failed Protests on Wild Public Networks

The Case of Dalian's Anti-PX Protests

Elizabeth Brunner

I n August 2011, a typhoon hit Dalian, a wealthy coastal town in China known for its beautiful coastline, parks, and beaches. The waves of the storm breached the large dikes built to protect the Fujia Dahua PX plant, located in the Dagu Mountain Petrochemical Industrial Park, twenty kilometers northeast of the city center, which produces 700,000 tons of para-Xylene (PX) per year.[1] According to some reports, the storm damaged the holding tanks, causing them to leak dangerous chemicals into the water. According to other reports, people were evacuated only as a precautionary measure, but the factory remained intact.[2] Though largely quashed in state-sanctioned media, reports and rumors of the damage spread quickly over social media networks, stoking anger across the city. Prior to the event, many residents were unaware that the plant existed, as officials did not go through the proper approval process for the plant, which requires such projects to be announced to the people and put forward for a period of public comment. As outrage over the potential dangers caused by the accident grew, citizens began organizing a resistance on popular social media platforms such as Baidu's *tieba*, Renren, and Weibo.

In the course of a week, word of the planned protests spread and over 10,000 people (with some reports estimating crowds of upwards of 70,000) showed up in

Dalian's People's Square to demand the plant be moved to a safer location.[3] People gathered in front of government buildings, holding signs demanding that "PX get out!," wearing gas masks, chanting, and singing. Only hours after the demonstrations began, the mayor stood atop a police van and announced that the city would move the plant off their coastline. His impromptu speech was caught on video by the sea of smartphones that surrounded him and posted to social media in what appeared to be a triumph for the people. The local paper featured front-page coverage of the protests the following day. In talking with residents of Dalian several years after the event, however, it quickly became apparent that the mayor's promise had not been fulfilled. Some Dalian residents reported that the plant was not shut down for even one day and that it definitely was never moved, as promised.[4] Their mayor had lied to them—to all the people who gathered in the square and to all the people who watched the events unfold online.

This betrayal was a huge disappointment for those who engaged in a very risky form of protest that left some participants jailed.[5] Though China has since changed its policies to demand that local officials prioritize the environment, during the decade of protests between 2007 and 2017, some local officials used false promises as a means to disperse the crowds.[6] These local leaders, perhaps fearing national embarrassment and investigations into whether they were following environmental protocols, attempted to quell environmental protests by publicly acquiescing to the people's demands, but then privately continuing with the contested project.[7] Such protests can be written off as failures because they did not achieve their instrumental goal—to relocate or shut down a polluting factory. Yet, they still function as important case studies that warrant sustained attention when we consider the wider range of green communication practices in China and beyond.

As Polletta argues, participating in activism can help to build networks, develop innovative tactics, and formulate ways to hold leaders accountable, even as they fall short of their goals.[8] The Dalian anti-PX protests thus provide a significant case study illustrating how political changes occur beyond a protest's instrumental success or failure.[9] Of particular interest to me is addressing these protests within the context of China's larger environmental movement by attending to the ways they contributed to altering relationships, inciting advocacy efforts, generating energy around a movement, offering a model as well as lessons for future protests, and shifting affective valences toward a particular issue or topic.[10] In this chapter, then, I contribute to the growing body of literature on how environmental advocacy functions.[11]

Protests have been an important component of China's environmental movement over the past decade. Xiamen's 2007 anti-PX protests are often viewed as an important success and as the beginning of a wave of protests across the country made possible by blogs, bulletin board systems (BBSs), and later social media platforms, largely because they resulted in the relocation of the plant. Xiamen's success inspired a wave of similar protests, including Dalian's 2011 protests. Still, the protests in Dalian were unique in that the residents were protesting against an existing plant and asking that it be relocated, which would have been costly and required halting PX production during a nationwide PX shortage. Thus, the protestors confronted unique barriers that the Xiamen and subsequent environmental protests—in Ningbo, Maoming, Chengdu, Shanghai, and other cities—did not face. Despite the failure of the Dalian protests to relocate the plant, reports from other cities that staged PX protests repeatedly mention the protests in Dalian and Xiamen as paving the way for these subsequent mobilizations, which is why Dalian serves as the case study for this chapter.[12]

In order to trace the ways the Dalian (and subsequent) protests helped to feed off the energy from the Xiamen protests and inspire others to take measures to safeguard their own local environments, I deploy concepts and methods capable of tracing change and that consider the nuances of the movement as well as the role of social media in these uprisings, for much of the (mis)information that provoked these protests circulated on social media. Thus, in this essay, I identify the repercussions, changes in relationships, and renegotiations that occur in the aftermath of a protest event outside of instrumental success.[13] As will be discussed below, protests such as the one in Dalian helped to move what Michael Calvin McGee terms *the social*, as they increased distrust in local officials, damaged feelings of nationalism, strengthened networks of citizens, and (along with other protests) served as inspiration for other cities facing similar threats to the environment.

To track the movements of "the social" in many directions, I examined over fifty news reports in both Chinese and English news outlets; looked at hundreds of posts on social media platforms and Internet forums such as Baidu's *tieba*, QQ, *tianya*, and numerous other blogs and BBSs; and drew from over twenty interviews with both people who witnessed the protests in Dalian and residents of cities involved in subsequent PX protests during my six months of fieldwork in China.[14] Doing so allows me to map how information moves across platforms, incites outrage, and creates important on-the-ground connections essential to understanding how, if at all, the social is changing and in what directions it is moving.

While I focus on protests in this chapter, they play just one role in raising awareness about and involvement in environmental protection efforts. By focusing on a specific case study of one form of environmentalism in China, this research contributes to the larger portrait of environmental efforts discussed in other chapters in this book, which together help to articulate the complexity of China's environmental situation and efforts. As is the case in the United States, environmentalism in China has been provoked, at least in part, by crises, which often lead to protests. Thus, protests are an important component of this period of China's shifting attitudes and practices about environmentalism, especially as the country has been shifting to implement systems that encourage care, protection, and preservation. In short, the Dalian anti-PX protests are a potent case study of the collision of increased environmental awareness and rapidly expanding and vibrant online cultures.

Tracing Green Social Change

While studying artifacts in the form of speeches and protest events deemed successes is warranted and necessary, incorporating failures into academic discussions can enhance existing theories of social movements.[15] For example, Darrel Enck-Wanzer reminds rhetorical scholars, in his analysis of the Young Lords Organization's garbage offensive in New York City, that "While the practical goal of rhetoric may be to persuade people to act in one way or another, instrumental 'success' may not be the best criterion on which to base our judgments."[16] Likewise, Phaedra C. Pezzullo argues in her essay on boycotts and buycotts led by environmental justice organizations that it is "inadequate to judge them as 'positive wins' or 'negative losses,'" and instead, scholars must "consider the possibilities for change enabled by specific campaigns within specific contexts."[17] As Wanzer, Pezzullo, and others argue, even when social movements do not achieve their immediate goals, scholars can still benefit from following the diverse, unintended, and unexpected repercussions of resistance.

This move for a more robust understanding of social movements' impacts also follows Michael McGee's call for social-movement scholars to trace the multiple repercussions of protests and, in doing so, to chart the *movement of the social*.[18] Treating protest as a force that disrupts relationships and drives renegotiations in a variety of realms is important work and opens up space for scholars to examine how protests—whether labeled as successes or failures—make possible new ways of being in the world while precluding others. The success of protests in Xiamen, for example, made possible the idea that protestors in Dalian could also shut down a PX factory. It provided hope for the thousands of people who joined together in the streets believing they could create change.

With the advent of social media and its integration into social movements, scholars taking a networked approach to studying activism are increasingly looking to movements outside the success/failure binary. For example, Kevin DeLuca, Sean Lawson, and Ye Sun's study of the Occupy Wall Street (OWS) protests eschews the binary, instead tracing the repercussions the protests had on national conversations. The protests' repercussions, thus, become the focus of the study. They found that "in a mere few weeks, OWS changed the national conversation despite the initial neglect and dismissive framing by traditional mass media organizations."[19] Thomas Poell, Jeroen de Kloet, and Guohua Zeng also discuss the impacts of online political contention by turning to the Huili picture scandal of June 2011 and the controversy surrounding rally racer and novelist Han Han in December 2011, when both unfolded on the social media platform Weibo. Their essay, rather than focusing on the success of an online campaign, or the lack thereof, "examines how Sina Weibo's particular technological features, its user cultures and self-censorship practices, as well as the occasional government interventions, mutually articulate each other," thereby offering scholars "insight into how new publics are constituted and how symbolic reconfigurations unfold."[20]

These changes in how scholars examine protest are largely motivated by the increasingly panmediated environment that impacts the size, scale, spread, movement, and trajectories of protests.[21] China's media environment has changed dramatically in recent years via the widespread adoption of social media. In 2018, China already had 616 million social media users according to Statista, which is almost triple the 211 million social-network users in the United States. Furthermore, social media apps such as Weixin (or WeChat) provide users with an array of tools that dwarf US-based apps such as Facebook and Twitter. Importantly, social media offer scholars new tools to track change, as they leave a massive network of traces in their wake in the form of posts, blogs, images, and video. Scholars in both the United States and China today can shift (and are shifting) the focus of social-movement studies from singular texts to complex protest ecologies, situated in particular cultural contexts.

Tracing Wild Public Networks

This essay builds upon existing research about social movements and public advocacy by applying the concept of *wild public networks* to the case study of Dalian.[22] This concept is born from the overlap between media theory and social-movement studies. By studying environmental advocacy through this term, I seek to better explicate the ways places, people, and media intertwine to move people to act, engage, seek information, create alliances, spread rumors, and disengage. Wild public networks meld Kevin DeLuca's concept of (*wild*) *public screens* with Joshua Ewalt et al.'s *networked* public screens and Bruno Latour's networked approach.[23] DeLuca's notion of the public screen is, in part, a reaction to and replacement for Habermas's concept of the public sphere. Unlike the public sphere, which emphasizes rational arguments delivered in person, public screens "highlight dissemination, images, hypermediacy, spectacular publicity, cacophony, immersion, distraction, and dissent" as well as glances, images, and panmediation.[24] Public screens in the form of newspapers, billboards, television screens, and smartphones are a reaction to media infiltrating every sector of society—from politics to history to activism. *Wild* public screens, according to DeLuca, Ye Sun, and Jennifer Peeples, are defined as "public screens full of risk, without protection, without guarantees." In China, "there are no guarantees of domesticating protection, where the ritualized performances of the public sphere are absent, but there are risky and powerful conversations and protests and activism." Wild public screens, then, are "characterized by *arrangiasti* (making do in the civic spaces of the world)."[25]

Joshua Ewalt, Jessy Ohl, and Damien Pfister's work emphasizes the spaces of dissent that produce gentler and quieter images in order to tend to the "ability of individuals to create their own networked media ecologies."[26] Their concept of networked public screens seeks to account for communication outside of dramatic and violent events to better capture how mundane conversations sustain and grow networks. Melding these two concepts together offers a nuanced portrait of the ways media and activism inform one another and make possible new networks of people and information that scholars can follow.

To this notion of *wild public networks*, I add a Latourian twist, which stresses not screens but networks and relationships and the importance of tending to these in-between spaces. To look at the protests through the concept of wild public networks draws attention to relationships over individuals, to the movement of information over content, and to affective appeals over rational arguments. These movements are now traceable and can offer scholars a more complex understanding of social *movement*. As Latour argues, "information technologies allow us to trace the associations in a way that was impossible before. Not because they subvert the old concrete 'humane' society . . . [but because] they make *visible* what before was only present virtually."[27] In other words, we can trace the flows of information, the connections that arise, and the images and videos that generate the most attention through social media, thus tracing the flow of ideas about green communication.

My deployment of this concept also addresses Pezzullo's call in the introduction to this volume to offer new concepts beyond the public sphere, for the idea of public spheres may not be an apt way of describing the role of publics in creating environmental change, in particular, or social change in China. The changing mediascape has made the idea of a sphere problematic, as social media extend horizontally in rhizomatic arrangements that look nothing like a sphere. Furthermore, when applied to the case of China, the concept of the public sphere has been widely debated and criticized, with scholars often arguing that its usefulness is limited, and calling for a new model that better represents the complexities of Chinese culture.[28] Thus, wild public networks offer an alternative concept that may better help to address green communication in contemporary China.

When applied to the Dalian 2011 protests, the concept of wild public networks opens up space to trace these networks via changes in relationships, renegotiations between citizens and their government, and new pathways formed by these changes. In doing so, it is important to note that the changes taking place are "not done under the full control of consciousness; action should rather be felt as a node, a knot, and a conglomerate of many surprising sets of agencies that have to be slowly disentangled."[29] By following the traces left on social media and technological devices, we can begin to map wild public networks. Indeed, the notion of wild public networks advocates for "an ontological shift from studying things to studying connections among things," so that scholars can better map "social movement" regardless of a protest's instrumental outcome.[30] This method looks to the traces of change left in the form of comments and images while acknowledging the temporality of activism. Within this approach, citizens become activists who then become changed actors in new publics, meaning activism can shift relationships, assemblages, networks, and pathways that connect citizens to other citizens, to smartphones, to apps, to land, and to political leaders.

Stacking Networks in The People's Favor

The protests in Dalian were deeply dependent upon social media networks, in part, because the protests were organized and took place within a restricted media environment. Since China's mass media is regulated by the state, officials heavily dictate what issues are addressed and how.[31] While these restrictions can be circumvented, doing so is risky for individual journalists, their superiors, and coworkers. Since the rise of social media, citizens have become more acutely aware that government directives guide state-controlled media, which decreases the state media's credibility.[32] Thus, I argue that people turn to the wild networks of social media, where they can connect and share (mis)information. Crisscrossing webs of relationships made possible by users engaging on Chinese social media platforms, including Weibo, Weixin, Renren, and QQ, twist and transform dominant cultural, technological, and commercial imperatives.

By 2011, when the Dalian protests took place, millions of people in China were online. A plethora of social media platforms such as QQ, Renren, Baidu's *tieba*, and Weibo had been developed; smartphones were finding their way into more people's pockets; Chinese citizens were becoming more dependent upon online networks for information; and networks of users and devices were transforming into denser and more complex systems. In 2011, China had 485 million Internet users and 235 million social media users and boasted the world's biggest smartphone market with 900 million mobile phone users.[33] Registered QQ users reached 500 million (a number that indicates some people had more than one account) and QZone was China's biggest traffic distribution platform.[34] In late 2011, registered Weibo accounts were upwards of 233 million, and the number of Renren accounts sat at approximately 137 million, meaning that people were using multiple platforms to connect.[35] Simply put, "microblogs in China [were] mushrooming across the web, including variations from Xinhua, Baidu, Netease, and Shanda's Tuita."[36]

A lack of trust in mainstream media refocused people's attention on online networks, where they went for news and commentary. Social media were deemed more credible because they provided reports in real time, content was comparatively unfiltered, and critiques of the government abounded. Though platforms were censored, they were far less restricted than state-sanctioned media. In 2011, according to a report by the Communication University of China in Beijing, "Microblogs like Sina Weibo [were] the third favorite online news source in China."[37] The same report stated that improper behavior or remarks made by government officials accounted for 70 percent of trending topics.[38] In short, political scandals drew interest and users to social media at the same time they increased skepticism about local officials. Netizens in Dalian and elsewhere had watched fellow social media users expose officials for corruption on Weibo.[39] Networks of common people commented, argued, and debated political issues in QZone groups. Wild public networks were and are home to information, rumors, dissent, and ideas that crisscrossed with an unstable environment saturated with distrust and concern in Dalian.

A Tangle of Connections

The protests that erupted in Dalian were fomented by a number of factors unique to the city, including an environmentally conscious public, political change, memories of previous environmental disasters, a nationwide PX shortage, and a growing culture of protest across China. Though historically Dalian had been known for "oil refining, chemical production, and ship building enterprises—and for pollution," after Bo Xilai took office as mayor of Dalian in 1993, Dalian's image changed.[40] In just five years, Dalian was designated a National Model City for environmental protection and was named China's Top Tourism City. Green economic growth was encouraged. For example, in 1996, Dalian rejected fourteen projects due to "potential pollution impacts and 206 factories were fined for failing to follow *santongshi*," a Chinese policy put in place to "ensure that environmental regulations are addressed at the planning, construction, and operation stages of project development."[41] These initiatives led to international recognition in the form of regional, national, and UN awards. Ultimately, the public nature of Bo's initiatives and open commitment to greening Dalian fostered important relationships between the people and their land, air, and water.

Bo's successor, Xia Deren, took Dalian in a very different direction. Xia refashioned Dalian into a petrochemical capital. Between 2008 and 2013 the city "dealt with huge oil leaks, massive industrial explosions, and all seafood [that came] out of its surrounding waters [was] polluted."[42] In particular, just a year before the typhoon hit Dalian, an oil storage-depot explosion spilled 1,500 tons of oil into the Yellow Sea. In the aftermath of these events, public concern over the environment grew. This change in attitude was apparent in conversations I had with people in Dalian in 2014, some of whom lamented the changes to Dalian's landscape. Designated green spaces had been turned into construction sites for hotels and apartment complexes. High rises now obscured the view of the ocean. The air quality was also deteriorating. In 2013, Greenpeace ranked Dalian as having the 57th worst air quality in China (though it has since improved tremendously, ranking 17th cleanest air in a 2018 survey of Chinese cities).[43] These changing relationships—between people and the air, citizens and their leaders, water and the fishing industry, green spaces and hotels—are important traces to attend to in mapping the wild public networks that extend out in many directions from the protest. Using a networked approach that tends to these details, "it is possible to trace more sturdy relations and discover more revealing patterns by finding a way to register the links between unstable and shifting frames of reference."[44]

A second factor that aided in fomenting protest was a nationwide shortage of PX, which was exacerbated by a growing culture of environmental protest in China. In 2005, the year the contract was signed to build the Fujia plant, China was already dealing with a massive shortage that forced it to import PX from Japan and Korea. While China's PX production was 2.23 million tons, its consumption exceeded this figure by 1.5 million tons.[45] The plant outside Dalian was seen as a way to dramatically decrease reliance on exports. PX is an important chemical in China's industrial chain and is used in the production of numerous polyesters, plastics, and other synthetic materials. Without PX, China cannot produce many of the products it consumes and exports as a rapidly developing country. This means, then, that Dalian's anti-PX protests coincided with a nationwide exigency regarding the industrial production chain, once again repeating a trope in which environmental issues confront economic growth.

To complicate matters, two years after the Dalian project was approved, in 2007, protests broke out in the coastal city of Xiamen against a proposed PX plant. In efforts to create a strong opposition case and garner the support of the Chinese People's Political Consultative Conference (CPPCC), Dr. Zhao Yufen, professor of chemical biology with dual appointments at Xiamen University and Tsinghua University, released information deeming PX to be a dangerous chemical. As the information moved from blogs to BBSs to casual conversations, it morphed and ran wild. On May 25, 2007, a text message began flying from one flip phone to the next declaring that allowing a PX plant to be built in Xiamen was akin to dropping a bomb on the city.[46] Outrage gathered across screens and streets into a *force majeure* that brought over ten thousand people together in opposition to the plant.[47] Xiamen leaders ultimately acquiesced to the people's demands and canceled plans for the project, eventually relocating it to the smaller town of Zhuangzhou.

The Xiamen demonstrations, which were the first to deploy cellphones and online forums, garnered widespread national and international coverage as well as intense attention on social media, stoking fear of PX across the country and ultimately leading other cities to also reject PX. Rumors spread across Weibo and other platforms that Xiamen's rejected plant was being relocated to Dalian, thereby creating a connection between the two projects that transmitted concerns over PX's safety. Though talk of the plant became a trending topic in Dalian, the government ignored citizen concerns, requests for information, and an open letter by reporter Lu Renzi in 2009.[48] Concern over the project thus waned until 2011, when the typhoon that breached the dikes of the fully constructed PX plant drew attention to it once again. People who were unaware of the plant's existence turned to wild public networks to find information where they learned that the Fujia Petrochemical Company—the company that built the PX plant—did not receive the mandatory environmental approval from the Environmental Protection Bureau until ten months *after* the plant had opened. Furthermore, local officials never shared plans for the plant with residents prior to construction, nor did they open up the issue for a period of public comment, which is also required by law. While "these serious breaches of process should themselves have led to severe sanctions, . . . thanks to local government support, the project quietly went ahead, out of public sight."[49] The typhoon acted as a catalyst, however, that rallied opposition and intensified fears of PX, thereby changing the ways people participated as citizens and inventing new publics.

A Storm Brews

As people began navigating wild public networks, in addition to copious amounts of information and rumors about PX and the potential dangers it posed, they discovered accounts of Xiamen's protests, which were touted as a monumental (instrumental) "success," discussed widely on Internet forums and blogs. People with whom I spoke in Dalian were aware that protests had been an effective way to protect Xiamen's local environment, and learned that officials bowed to the demands of citizens in the face of a mass demonstration.[50] Access to these new flows of information opened up possibilities for reconstructing relationships between people and their government through protest. Indeed, rumors had stirred panic among the masses as they hopped from screen to screen. The state media, which had intended to report on the alleged spill, was not allowed near the plant and later received directives to pull the story. With no official media to which they could turn, people began participating in fear-induced conversations about a chemical spill many feared was real. Some people likened the spill to the Fukushima nuclear disaster (though this is an exaggeration). They read hyperbolic reports about the potential impacts of PX exposure. One post on Weibo read: "The power of the 1,000 missiles aimed at Taiwan is far less than that of the [anticipated] PX explosion." Echoing the Xiamen protestors, others compared PX to an atomic bomb. Still other users alleged PX caused leukemia and birth defects. Rumors spread rapidly. Dalian citizens traded links to stories and summarized their findings for others on social media. Networks of social media, friends, coworkers, and family were abuzz with chatter, and anger at officials and their silence drove much of the conversation. The flows of communication were characterized by wildness, and it was strengthening communication networks among the people at the same time that it created resentment and anger toward local leaders. Networks were shifting.

As fear and rumors spread, plans for a protest began coalescing over social media. According to interviewees, information about the protests was easy to access, and as the protests drew nearer, the government posted a warning that appeared on one of the BBSs used to organize the event, telling users not to participate in the protests. The people, cynical after watching their government repeatedly privilege industry over citizens, were willing to risk freedoms for a safer living environment. Relationships between the people, the environment, the government, and protest were changing. Then the energy crossed over from wild public networks into more traditional forms of action, as people showed up in the public square armed with banners, signs, smartphones, and anti-PX T-shirts. They marched, sat, and demanded the PX plant be moved for hours, until the mayor, who had been in office only two months at the time of the protests, stood atop a police van and offered a promise that would never be fulfilled. Though the Dalian Municipality reportedly sent letters to every resident affirming the plant would be shut down, many think production was never halted. The nationwide PX shortage made it too valuable a chemical, and relocating the plant was far too costly a project to realize.

Though the protests did not result in the plant being moved, they did forge pathways by which people communicated with their government and offered new enactments of citizenship for the people of Dalian and beyond. This failure made evident problems in current relationships by highlighting the need for transparency between citizens and their government, the lack of public trust in officials, and the level of investment people had in their living environment. The Dalian anti-PX protests thus demonstrate how a crisis created new grounds for imagining new forms of green communication.

Tracing the Movement of the Social

The protests indicate the movement of the social, in that people chose protest as a vehicle to air their grievances. The event instigated further changes in relationships. As has been established, the protests in Dalian were ultimately an instrumental failure. The plant continues to operate at the time of this writing (2018). This does not, however, mean there were no repercussions. The impacts of the protest reverberated beyond its failure. Scholars must consider and trace the unanticipated movements of the social via the wild public networks over which they erupt. As Mike Gunter Jr. argues, "To label Dalian a [Not In My Backyard] NIMBY failure, using Western democratization-school standards of civil society activity, is perhaps . . . a rush to judgment . . . Much more interesting than the wins and losses of continuing plant operation is the process by which environmental activists let their displeasure be known—and the process by which governmental actors reacted." If scholars consider the repercussions of the protests, one can see that they did, indeed, extend beyond whether it counts as a win or a loss. Gunter further asserts that "Dalian highlights the degree to which NIMBY events are much more complicated than the black-and-white rhetoric of victory and defeat."[51] I agree with and extend Gunter's argument, concluding that the protests (1) provoked a further lack of trust in officials, (2) damaged people's pride in their city, (3) increased networks among residents, (4) inspired people to protest in other cities, and (5) compelled leaders to respond to China's environmental crises. In the following pages, I elaborate on these five arguments, thus showing how the instrumental "failure" in Dalian helped to produce long-term growth in wild public networks.

Growing Distrust

Prior to the protests, a lack of trust in the local government was building. The environmental degradation Dalian suffered under Xia and the lack of transparency from government officials caused many citizens to lose trust in local officials. Thus, when the PX plant was allegedly damaged during the typhoon, few believed the government's official statement that the plant had not sustained damage. The waning trust in officials was apparent in protestors' comments after the protests. One protestor told reporters, "Even if there was contamination, the government would restrict the news."[52] Another told reporters he protested because the people of Dalian "know that the typhoon caused some leak of poisonous chemicals from the PX project and we are all worrying about it because it is a threat to our life."[53] The protestors had learned to rely on social media for information because little to no official coverage existed and many citizens did not trust what was reported. The relationship between people and their government was problematized and strained. One interviewee expressed frustration with the lack of mass-media coverage, stating that "Many of the major issues surrounding us we don't find out about on domestic websites, but from foreign websites. We actually live in a world without truth, at least for the Chinese." Another interviewee reported that the protests "impacted people's confidence in their government," thereby hurting the relationship between the government and its citizens.

During the protests, mistrust, which flew across social media platforms and Internet forums in the form of posts and pictures, strengthened. An image of a group of people collectively giving the middle finger to the government offices circulated widely. A Chinese Twitter user who was cross-posting from Weibo expressed deep dissatisfaction with the way the incident was handled.[54] Still others compared the Dalian issue to corruption cases, including the highly publicized Wenzhou train crash that resulted in the death of over forty people and more than two hundred injuries.[55] Yet another user commented that though the officials announced the PX plant would be moved, the "government's trustworthiness" was nonexistent, which would provoke speculation.[56] As the protests continued, the posts questioning authorities grew. If the government thinks PX is safe, posted @tatamama, then they should build the government offices next to the plant.[57]

This mistrust is also apparent after the protests, as people discovered that their government had lied. The outcome, the protestors said, hurt the mayor's image and caused people to further lose confidence in the government. People with whom

I spoke told me that the current mayor judges his worthiness on his economic development, not his environmental safety record, which is why the plant was permitted to be built. The Fujia PX plant was and is an enormous revenue generator for the city. One official from the Dalian propaganda office was quoted as saying, "We need to consider the profit of business. It takes time to move the plant. If the production is halted before the relocation, the business will be bankrupt."[58]

Waning Nationalism

In addition to exacerbating the lack of trust in officials to protect the people, the Dalian protestors said the event and the violent nature of it also damaged their enthusiasm for their country and city. The disappointment in Dalian was evident during my research trip. As was previously described, people were quick to criticize the current mayor, and to decry the changes he initiated, at the same time they expressed a longing for the city they previously knew. Public approval of the new mayor was low by many measures, and his unwillingness to listen to the people he ostensibly served did not gain him any favor with residents.

The people's relationships with the government, each other, and the environment had been altered by the protest and its failure. One protestor commented: "This protest has little effect on the factory, but it has a great effect on society ... It impacted people's confidence in their government."[59] Not only were the participants distraught over the outcome, but they were also afraid that the violence and instrumental failure would create despair for the youth who participated. This younger generation was next in line to protect Dalian, to fight for their home. If this was how the government reacted to protest—which is a constitutional right—this next generation might be more reticent to participate in uprisings. Perhaps, they conjectured, protests were not an effective way to voice their opinions. Thus, people were driven to find alternative forms to express their concerns about a local government they no longer trusted.

Increased Solidarity

Another impact of the protest was that it created new networks of support by raising awareness about the issues at hand, which, in turn, increased support for the protestors that extended far from the streets. Some people who did protest remain united by their shared experience. Those who were injured and jailed share special connections because they suffered similarly. Those not directly involved in the demonstrations (perhaps because it was too risky) showed support by buying water and snacks for the people holding signs and chanting. Others showed support in additional ways, including utilizing their *guanxi* to help friends get out of jail. One protestor whose friend was injured during the protests and subsequently hospitalized was very concerned about taking time off from work to go visit him; he was reticent to reveal to his supervisor why his friend had been injured. Not only did his supervisor give him leave from work to go see his friend, but he also covered the hospital fees. The bonds created between these people highlight how suffering and injustice can link people in important and consequential ways.

Inspiring Action Afar

Repercussions pulsated beyond Dalian's jurisdiction. While the people of Dalian dealt with the disappointment, many people in Ningbo, Kunming, Maoming, Hangzhou, Qidong, Quanzhou, and other cities never followed up on reports that the protests failed, which were not widely covered in the mass media. Many outsiders believed the protests were a success after hearing about them on social media and through group chats and searching the Internet for stories, and found in them, as well as the 2007 Xiamen protests, motivation to stage their own oppositions. Each of these aforementioned cities staged large-scale protests that were, in part, driven by the information people gathered online. When searching for PX on the Internet in China, one is confronted with tens of thousands of stories about not only PX but also the protests it has incited. As described previously, rumors, information, exaggerations, and anger continue to circulate across social media networks. For example, after the 2012 anti-PX protest in Ningbo, the newspaper *Qilu Evening News* (*Qilu Wanbao*) reported that "Because of the problem with PX, this same mass incident have [*sic*] already occurred in Xiamen, Dalian, and other cities, and there is no way that Ningbo didn't know about these incidents in other cities."[60] Social media and Internet search engines made searching for "PX" easy. The results, inevitably, included information about what happened in Xiamen, and then Dalian, as well as rumors and partial stories. Many may have seen the video of the mayor of Dalian telling the people he would shut down the plant, but not the story about the plant not shutting down. Or, they may have read small sections of posts that proclaimed victory, but not the results that demonstrated the protests were an instrumental failure.

For those who were aware that the plant in Dalian was not shut down, this lesson gave protestors reason to continue protesting even after local officials made promises. Those in Ningbo persistently protested "even after the authorities pledged to halt the PX project, in part because suspicion of the government runs so high. 'We don't trust them at all; we think [their promise] is a stalling tactic,'" said one participant.[61] This distrust is cultivated through stories that reveal corruption, lies, and the valuing of GDP over environmental safety, such as the one in Dalian.

In 2013, the people of Kunming staged two separate protests in a single month against a planned PX plant. They, too, went online to find out about PX, but as Li Bo, director of China's first ENGO, Friends of Nature, put it, the people of Kunming "had a lot of difficulty getting satisfactory information" about PX. Li went on to say that "A lot of worries and doubts have accumulated [around PX], which is more or less what happened with the previous PX projects in Dalian and Ningbo."[62] After witnessing the protests in Kunming, the authorities in Chengdu "appear[ed] nervous that large-scale environmental rallies could spread there because of public anger over a PX plant in Pengzhou, on the city's outskirts." And they did. The people in Chengdu rose up against their government, demanding that "PX get out!" People were finding that the government in place was broken. As Lubman writes, "In order for public participation to work, both officials and citizens must be educated and trained on its mechanics, such as the conduct of public hearings. This is no easy matter, since local officials often neither seek nor heed public sentiment and the public, in turn, has often grown doubtful about officials' motivation."[63]

This is a problem across China and, until it is addressed, will likely continue to contribute to further clashes and citizens' use of any means necessary to ensure a safe environment for themselves and their children. Dalian is not responsible for all of this unrest, but protests there are worth recalling as part of a larger environmental movement that has incited important change in China and launched environmental concerns into the public awareness. Indeed, repeated unsanctioned protests have indicated to local and national officials in China that people are becoming increasingly aware of environmental concerns and are willing to protest to protect their bodies from dangerous chemicals.

Influencing Political Discourse

As the protests in Dalian and subsequent protests in other cities garnered increasing attention locally, nationally, and globally, leaders were forced to address the issues at hand. China was facing a growing number of environmental protests and increased unrest.[64] In July 2012, then President Hu Jintao, in a speech addressed to CPC leaders and officials, told his audience that "people's demands for a better life and expectations for prompt solutions to prominent social problems were increasing."[65] Surveys of people's concerns show President Hu was correct. In November 2012, *China Daily* found that environmental degradation ranked fourth on over half of the respondents' list of concerns, only after the wealth gap, corruption, and the power of vested interests. Growing anxieties and unrest may have been part of the motivation behind the announcement made during the Communist Party conference in 2012, in which central Party officials "added the environment to the four 'platforms'—basic beliefs that define what the party stands for."[66] Additionally, the government increased its budget for *weiwen*, or "stability maintenance," so much so that it spent more on internal security than defense beginning in 2010.[67] This disparity "shows the party is more concerned about the potential risks of destabilization coming from inside the country than outside."[68] Citizens were helping to change their relationship with the government at the highest level, and this forced a renegotiation of relationships among the people, environmental laws, top-level officials, and government spending. More recently, in 2017, China publicly announced its plans to become the world's green leader and is taking steps to reduce pollution that are working. Air quality in cities like Beijing is rapidly improving, with PM2.5 levels in Beijing down 54 percent in 2018, according to Greenpeace.[69] Many places are working aggressively to transition from coal to cleaner forms of energy such as natural gas and renewable resources. These changes are remarkable in both scale and speed.

As Latour argues, "a society needs new associations in order to persist in its existence."[70] Each protest changes relations between the people and their government, not only in a particular location, but across China. Each new protest adds force, creates a precedent for change, and inspires other people to stand up for their rights and ask for increased transparency, a better system of checks and balances, reduced corruption, and a safe living environment.

Conclusion

The Dalian case study makes two contributions to environmental communication research in China, the United States, and beyond. The first is a move toward incorporating instrumental failures into research. In a panmediated society, scholars have the ability to more readily trace change, especially as it occurs across distance and time. Tracing this change is essential to better understanding how and why social movement occurs in complex protest ecologies. The second contributes to the expansion of literature on wild public networks, shifting the focus from tracing social movement via "success" to investigating the changes in relationships brought about by rhetorical interruptions and inventions via images, words, and bodies. As my findings indicate, civic engagement is changing in China; people are becoming more involved and are trying to determine how they can renegotiate relationships with their officials to create more productive exchanges and better environmental governance. Regardless of the instrumental success or failure of protests, these events continue to enhance and shift relationships in dramatic ways.

This essay also proffers new concepts that contribute to the public sphere debate, which will aid both US-based and China-based scholars in studying social movement, especially as it interacts with social media. Wild public networks offer an alternative model to the public sphere because the concept acknowledges the wildness of public dialogue as it traverses physical spaces and Internet platforms, creating myriad crisscrossing flows of information, rumors, dissent, concern, panic, and confusion. These networks pulsate, grow, contract, and are energized by people as they interact. Wild public networks assume neither democracy nor rational debate. They are rife with movement, contradictions, censorship, confusion, and affective pleas. Wild public networks are concerned with changing relationships and connections rather than specific arguments or actors.

US models of the public sphere often emphasize freedom, rationality, democracy, and individual actors, which is why authors such as Gu Xin have leveraged a critique of its application to the Chinese case, asking scholars to attend to the important differences in China that make easy application of the theory dubious. Indeed, China's political context is far different than the United States'. For example, to assume that China is moving toward democracy is problematic, as democracy is not inevitable even as public participation and protest increase. Further, the Western model of rationality, which is a complex system of thought often presented as natural or obvious, is not universally present across cultures, nor is it accepted as the ideal form of problem-solving. Ethics, which focus more on relationships and hierarchies, often act as an important paradigm for problem-solving in China. Thus, wild public networks, by focusing more on relationships, shifting hierarchies, affect, and wildness, can better adapt to the China context.

The 2011 Dalian protests produced repercussions that stoked the fire of hope necessary for the environmental movement to continue, even when faced with setbacks. They can also, however, lead to further setbacks. For example, these protests provoked increased distrust in officials, which has the potential to improve government conduct, yet at the same time the protests have the potential to sow the seeds of pessimism, even as they can help to motivate the people to strengthen new networks and develop new forms of resistance and advocacy. These associations could lead to the development of local environmental groups that offer alternative and credible sources of information when ecological crises arise. Though many people lost pride in a city they once felt honored to call home, those distraught about the changes hold knowledge and memories of what a green city could be. This provides space to reimagine how to return Dalian to its former post as a model environmental city. Beyond its own city borders, Dalian serves as a warning and a beacon of hope. But other cities inspired by Dalian to protest against PX and other chemical plants must do more than protest: they must develop measures to ensure that the promises made by officials are kept. And as more protests occur, top officials must do more than respond—they must take action to protect public health. Though disappointment hangs in the air in Dalian, so too does the potential for alternate forms of civic engagement fostered by new relationships across wild public networks.

NOTES

1. Andreas Rinke, "China's Dalian Fujia Plant Runs Normally Despite Shutdown Order," *Reuters*, August 15, 2011.

2. Sina, "Dalian Chemical Plant's Anti-Tidal Wave Dam Continued: The Accident," Sina.com, August 10, 2011, http://news.sina.com.cn/c/2011-08-10/083222966273.shtml; Hao Jin, "20 Meter Waves Breach the Dalian Seawall Threatening Highly Toxic Chemical Tanks," QQ News, August 9, 2011, https://news.qq.com/a/20110809/000122.htm.

3. Jonathan Watts, "Tens of Thousands Protest against PX Chemical Plant in Northern China," *The Guardian*, August 14, 2011.

4. As was reported by interviewees during my fieldwork.

5. Yat-yiu Fung, "Dalian Tries PX Activists," Radio Free Asia, January 22, 2013, https://www. refworld.org/docid/511ce44528.html.

6. Samantha Hoffman and Jonathan Sullivan, "Environmental Protests Expose Weakness in China's Leadership," *Forbes Asia*, June 22, 2015.

7. Dan Wang, "How Credible Are China's Environmental Promises?," Radio Free Asia, July 7, 2016, https://www.rfa.org/english/commentaries/promises-07072016112523.html.

8. Francesca Polletta, *Freedom Is an Endless Meeting: Democracy in American Social Movements* (Chicago: University of Chicago Press, 2004).

9. Michael Calvin McGee, "'Social Movement': Phenomenon or Meaning?," *Central States Speech Journal* 31, no. 4 (December 1980): 233–244.

10. This approach is informed by Jun Liu, "Digital Media, Cycle of Contention, and Sustainability of Environmental Activism: The Case of Anti-PX Protests in China," *Mass Communication and Society* 19, no. 5 (September 2, 2016): 604–625.

11. Phaedra C. Pezzullo, "Performing Critical Interruptions: Stories, Rhetorical Invention, and the Environmental Justice Movement," *Western Journal of Communication* 65, no. 1 (Winter 2001): 1–25; Charles J. Stewart, Craig Allen Smith, and Robert E. Denton, *Persuasion and Social Movements*, 6th ed. (Long Grove, IL: Waveland Press, 2012).

12. Anonymous, "Renrenningbo, mianchaodahai, chunnuanhuakai (People of Ningbo, face the Korean Sea, in the warmth of spring flowers will bloom), *Minjian Yuwen* (*Folk Chinese*) (blog), October 29, 2012, http://bbs.tianya.cn/post-free-2849477-1.shtml; Andrew Jacobs, "Rare Protest in China against Uranium Plant Draws Hundreds," *New York Times*, July 12, 2013; Luna Lin, "Residents in Maoming Protest against PX Production," *China Dialogue* (blog), March 4, 2014, https://www.chinadialogue.net/blog/6878- Residents-in-Maoming-protest-against-PX-production/en; People's Daily, "People's Daily Comment on People's Resistance to PX Project: 'Walking' Is the Best Way," QQ, May 8, 2013, http://news.qq.com/a/20130508/000520.htm; Shannon Tiezzi, "Maoming Protests Continue in Southern China," *The Diplomat*, April 5, 2014, http://thediplomat.com; Jiajun Wang, "Why Is PX 'More and More Dangerous,'" *Phoenix Weekly* (blog), June 29, 2015, http://blog.sina.com.cn/s/blog_4b8bd1450102vm4r.html.

13. Elizabeth Brunner, "Wild Public Networks and Affective Movements in China: Environmental Activism, Social Media, and Protest in Maoming," *Journal of Communication* 67, no. 5 (September 7, 2017): 665–677.

14. All research was conducted under ethical guidelines per the University of Utah Institutional Review Board (IRB_00064671).

15. Jan Golinski, *Making Natural Knowledge: Constructivism and the History of Science* (Chicago: University of Chicago Press, 2005); Thomas Parke Hughes, *Rescuing*

Prometheus (New York: Vintage Books, 2000).

16. Darrel Enck-Wanzer, "Trashing the System: Social Movement, Intersectional Rhetoric, and Collective Agency in the Young Lords Organization's Garbage Offensive," *Quarterly Journal of Speech* 92, no. 2 (May 2006): 189.

17. Phaedra C. Pezzullo, "Contextualizing Boycotts and Buycotts: The Impure Politics of Consumer-Based Advocacy in an Age of Global Ecological Crises," *Communication and Critical/Cultural Studies* 8, no. 2 (June 2011): 139.

18. McGee, "Social Movement."

19. Kevin M. DeLuca, Sean Lawson, and Ye Sun, "Occupy Wall Street on the Public Screens of Social Media: The Many Framings of the Birth of a Protest Movement," *Communication, Culture & Critique* 5, no. 4 (2012): 483–509, 484.

20. Thomas Poell, Jeroen de Kloet, and Guohua Zeng, "Will the Real Weibo Please Stand Up? Chinese Online Contention and Actor-Network Theory," *Chinese Journal of Communication* 7, no. 1 (January 2, 2014): 2, 14.

21. Elizabeth Brunner and Kevin M. DeLuca, "The Argumentative Force of Image Networks: Greenpeace's Panmediated Global Detox Campaign," *Argumentation and Advocacy* 52, no. 4 (2016): 281–299.

22. Brunner, "Wild Public Networks and Affective Movements in China."

23. Bruno Latour, *Reassembling the Social: An Introduction to Actor-Network-Theory* (New York: Oxford University Press, 2007).

24. Kevin M. DeLuca, Ye Sun, and Jennifer Peeples, "Wild Public Screens and Image Events from Seattle to China," in *Transnational Protests and the Media*, ed. Simon Cottle and Libby Lester (Global Crises and the Media, vol. 10) (New York: Peter Lang, 2011), 154.

25. DeLuca, Sun, and Peeples, "Wild Public Screens," 154.

26. J. P. Ewalt, J. J. Ohl, and D. S. Pfister, "Activism, Deliberation, and Networked Public Screens: Rhetorical Scenes from the Occupy Moment in Lincoln, Nebraska (Part 1)," *Cultural Studies & Critical Methodologies* 13, no. 3 (June 1, 2013): 187.

27. Latour, *Reassembling the Social*, 207.

28. Guobin Yang's influential work, along with a number of different coauthors, helped to draw attention to the "green public sphere" emerging in the 2000s wherein environmental discourses were intersecting with new publics and a fast-changing media environment. His work has informed numerous subsequent publications that consider the widespread adoption of social media in China to inform new concepts. As Jingfang Liu reminds us, the Internet alone does not and cannot bring about social change, but creative users who adopt it for environmental ends can bring about important change. On the debate, see Gu Xin, "Review Article: A Civil Society and Public Sphere in Post-Mao

China? An Overview of Western Publications," *China Information* 8, no. 3 (January 1, 1993): 38–52; Frederic Wakeman, "The Civil Society and Public Sphere Debate Western Reflections on Chinese Political Culture," *Modern China* 19, no. 2 (April 1, 1993): 108–138; J. Lagerkvist, "The Internet in China: Unlocking and Containing the Public Sphere" (Lund, Sweden: Dept. of East Asian Languages, Lund University, 2006); R. Madsen, "The Public Sphere, Civil Society and Moral Community: A Research Agenda for Contemporary China Studies," *Modern China* 19, no. 2 (April 1, 1993): 183–198; Mary Backus Rankin, "Some Observations on a Chinese Public Sphere," *Modern China* 19, no. 2 (April 1, 1993): 158–182.

29. Latour, *Reassembling the Social*, 44.

30. Brunner, "Wild Public Networks and Affective Movements in China."

31. Beina Xu and Eleanor Albert, "Media Censorship in China," *Council on Foreign Relations*, February 17, 2017, https://www.cfr.org/backgrounder/media-censorship-china. See also China Digital Media's archive of government directives.

32. China's media environment differs from the United States, where media are run independently and owned by businesses that operate separately from government. Despite their autonomy, this system is rife with its own set of problems.

33. Willis Wee, "China: 485M on Internet, 195M on Microblogs, 900M on Mobile," *Tech in Asia*, June 19, 2011, https://www.techinasia.com/china-internet-users-statistics.

34. This number—500 million—conflicts with the 235 million total users likely for two reasons. First, users will often register more than one account to a single name, either because they want to use more than one account or because they re-registered with a different username after forgetting an old username or password. Second, social media companies do not often account for this precisely because higher numbers equate to higher investment and profit.

35. Incitez China, "Tencent Reports 530M Qzone Users and 233M Weibo Users," *China Internet Watch*, August 12, 2011, http://chinainternetwatch.com/1229/tencent-q2-2011; Incitez China, "Renren Reached 137 Million Users in Q3 2011," China Internet Watch, November 16, 2011, http://chinainternetwatch.com/1298/renren-q3-2011.

36. Willis Wee, "Microblogs: The Third Favorite Online News Source in China," *Tech in Asia*, March 27, 2011, para. 4, https://www.techinasia.com/microblogs-the-third-favorite-online-news-source-in-china.

37. Wee, "China," para. 1.

38. Wee, "Microblogs."

39. Perry Link and Xiao Qiang, "From 'Fart People' to Citizens," *Journal of Democracy* 24, no. 1 (2013): 79–85; Jonathan Kaiman, "Chinese Anti-Corruption Drive Nets Official with 47

Mistresses," *The Guardian*, January 5, 2013; Maria Bondes and Günter Schucher, "Derailed Emotions: The Transformation of Claims and Targets during the Wenzhou Online Incident," *Information, Communication & Society* 17, no. 1 (January 2, 2014): 45–65.

40. Bo Xilai was convicted of corruption and sentenced to prison in 2013. Lisa Hoffman, "Governmental Rationalities of Environmental City-Building in Contemporary China," in *China's Governmentalities: Governing Change, Changing Government*, ed. Elaine Jeffreys (New York: Routledge, 2009), 108.

41. *Santongshi* translates to the "Three Synchronizations." National Research Council (US) et al., *Energy Futures and Urban Air Pollution Challenges for China and the United States* (Washington, DC: National Academies Press, 2008), 314.

42. East by Southeast, "Anatomy of a Protest: Kunming Citizens Voice Concern over Chemical Plant," *East by Southeast* (blog), May 12, 2013, para. 7, http://www.eastbysoutheast.com/anatomy-of-a-protest-kunming-citizens-voice-concern-over-chemical-plant.

43. Monica Tan, "Bad to Worse: Ranking 74 Chinese Cities by Air Pollution," *Greenpeace East Asia*, February 19, 2014, http://www.greenpeace.org/eastasia/news/blog/bad-to-worse-ranking-74-chinese-cities-by-air/blog/48181.

44. Latour, *Reassembling the Social*, 24.

45. Jing Li, "Dalia FujiaPX xingmu mingyun ji: yizuo gongchang yu yige chengshi de gushi" (Fate of Dalian Fujia PX project: the story of a factory and a city)," *Life Week*, August 26, 2011, http://www.lifeweek.com.cn/2011/0826/34723_2.shtml.

46. China.org.cn, "People vs. Chemical Plant," January 14, 2008, http://www.china.org.cn./environment/features_analyses/2008-01/16/content_1239503.htm.

47. Elizabeth Brunner, "Fragmented Arguments and Forces Majeure: The 2007 Protests in Xiamen, China," *Argumentation and Advocacy* 54, no. 4 (2018).

48. Yi Liu et al., "A Governance Network Perspective on Environmental Conflicts in China: Findings from the Dalian Paraxylene Conflict," *Policy Studies* 37, no. 4 (July 3, 2016): 314–331.

49. Tang Hao, "Public Storm in Dalian," *China Dialogue*, June 9, 2011, para. 4, https://www.chinadialogue.net/article/4511-Public-storm-in-Dalian.

50. This data was collected during six months of fieldwork in China.

51. Mike Gunter Jr., "The Dalian Chemical Plant Protest, Environmental Activism, and China's Developing Civil Society," in *Nimby Is Beautiful: Cases of Local Activism and Environmental Innovation around the World*, ed. Carol Hager and Mary Alice Haddad (New York: Berghahn Books, 2015), 152, 153.

52. Quoted in Watts, "Tens of Thousands Protest against PX Chemical Plant in Northern

China," para. 11.

53. "China Protest Closes Toxic Chemical Plant in Dalian," *BBC News*, August 14, 2016, para. 15, http://www.bbc.com/news/world-asia-pacific-14520438.

54. Bornanit, Twitter post, August 13, 2011.

55. @TinylightWong, Twitter post, August 13, 2011.

56. @WSHNHONG, Twitter post, August 14, 2011.

57. @tatamama, Twitter post, August 14, 2011.

58. Gunter Jr., "The Dalian Chemical Plant Protest," 152.

59. Dalian resident, interview, 2014.

60. Barry van Wyk, "Ningbo Will 'Resolutely Not Have the PX Project,'" *Danwei* (blog), October 29, 2012, para. 8, quoted in Daniele Brombal, *Proceedings of the XV East Asia Net Workshop* (Venice: Edizioni Ca' Foscari, 2015), 27.

61. Christina Larson, "Protests in China Get a Boost from Social Media," *Bloomberg*, October 29, 2012, para. 6.

62. Quoted in Jonathan Kaiman, "Chinese Protesters Take to Streets in Kunming over Plans for Chemical Plant," *The Guardian*, May 16, 2013, sec. World News, para. 18.

63. Stanley Lubman, "Vital Task for China's Next Leaders: Fix Environmental Protection," *Wall Street Journal*, October 19, 2012, para. 8.

64. Jun Liu, "Digital Media, Cycle of Contention, and Sustainability of Environmental Activism: The Case of Anti-PX Protests in China," *Mass Communication and Society* 19, no. 5 (September 2, 2016): 604–625; Larson, "Protests in China Get a Boost From Social Media"; Max Fisher, "How China Stays Stable Despite 500 Protests Every Day," *The Atlantic*, January 5, 2012.

65. Xin Zhou and Henry Sanderson, "Chinese Anger over Pollution Becomes Main Cause of Social Unrest," *Bloomberg News*, March 6, 2013, para. 6.

66. "The East Is Grey," *The Economist*, August 10, 2013, para. 22.

67. Ben Blanchard and John Ruwitch, "China Hikes Defense Budget, to Spend More on Internal Security," *Reuters*, March 5, 2013.

68. Quoted in ibid., para. 5.

69. Greenpeace East Asia, "PM2.5 in Beijing down 54%, but Nationwide Air Quality Improvements Slow as Coal Use Increases," Greenpeace East Asia, 2018, http://m. greenpeace.org/eastasia/high/press/releases/climate-energy/2018/PM25-in-Beijing-down-54-nationwide-air-quality-improvements-slow-as-coal-use-increases.

70. Latour, *Reassembling the Social*, 218.

The STEMing of Cinematic China

An Ecocritical Analysis of Resource Politics in Chinese and American Coproductions

Pietari Kääpä

I n the world of the film *Looper*, time travel has been discovered and outlawed.[1] The detection of murder has been advanced, and an industry has formed around sending individuals to the past to be "disappeared" prior to committing their violent acts. The story begins in 2044, when a hit man named "Joe" (Joseph Gordon-Levitt) from a Kansas City crime syndicate faces "closing his loop"—that is, killing his older self (Bruce Willis). Joe's present is set in a derelict city in the United States, filled by people who appear homeless, camping outside crumbling buildings, vacant lots, and outdated architecture. Due to the complex mechanics of time-travel paradoxes, his older self can anticipate the assassination attempt and escapes. Joe then flashes forward to 2074, where he has given up his criminal ways, settled into a relationship, and moved to Shanghai. His new setting features a gleaming, spectacularly polished high-tech landscape with towering skyscrapers and residents who appear to be thriving. This Chinese setting provides a direct contrast to the dilapidated state of the United States, which is seemingly on the brink of societal and environmental collapse. If the film is telling a story about global trends, it is clear which nation appears to have the greatest promise for future prosperity.

These visuals confirm an earlier comment from Joe's boss, Abe, who has been sent from the future to oversee operations. When Joe says he is learning French,

Abe responds: "Trust me, I am from the future, you need to learn Mandarin." This line reflects the production history of the film, which was originally to be partially shot in France but relocated to China. This shift onscreen also occurred as the US-based production company Endgame Entertainment partnered behind the screen with the China-based DMG Entertainment, founded by Dan Mint, Bing Wu, and Peter Xiao to help navigate China's often complex rules on film imports. Thus, both off-screen and on-screen, the action driving *Looper* has moved from the Old World of Euro-American dominance to China.

The year 2012, when *Looper* was released, marked a distinct advance in coproduction relationships, as we started to see both a significant increase in Chinese audiences and a more focused attempt to cater to them by Hollywood film studios. American trade-press publications like *Variety* started to publish articles on the potential significance of the market as a global powerhouse. The negotiations between governments led to an agreement signed in 2012, whereby China entered the parameters of WTO regulation. This agreement allowed for an increase in imported content through a partnership managed by the China Film Group Corporation (CFGC), which oversees any international co-venture arrangements like the Legendary/DWG partnership.[2]

Film scholar Deron Overpeck identifies this transformation as a part of a Chinese soft power strategy, where harnessing both content and professionalization strategies for the industry can lead to the wider dissemination of Chinese cultural and political values, a strategy used by the US media industry for decades in consolidating its now-dissipating global hegemony.[3] For film scholar Aynne Kokas, the key to understanding this increased level of investment in the media comes from infrastructural development:

> The PRC's twelfth five-year plan, released in 2011, identified the country's media industries as a major pillar for economic growth that should therefore receive central government support. Within the policy context of the PRC's twelfth five-year plan, Sino-US media are an extension of the PRC central government's role as a media industry stakeholder. The plan discusses the growth of media industries in terms of the development of hard infrastructure (physical spaces for industrial development) and soft infrastructure (workforce skills development).[4]

Thus, the key impetus for China's media and cultural policy framework is to generate growth in all parts of the domestic industry and to continue international expansion in terms of allowing more access to the markets and building domestic capacity—all while maintaining the political status quo.

Chinese collaborations with Hollywood studios make sense financially too. Many of these imported films continue to place high on the domestic box-office charts, and the coproduction projects bring in both artistic and capacity-building stimulation to the Chinese media economy. At the same time, they create an opportunity for Hollywood media conglomerates to extend their reach and production capabilities in China. Moreover, these cinematic collaborative ventures pose a positive incentive for the Chinese government to take advantage of the globally dominant role of Hollywood to export images of China to international audiences and for US producers to expand into a larger market. Major productions, from *Transformers 2* to *Iron Man 3*, and from *Pacific Rim* to *The Great Wall*, operate on the basis of agreements between the United States and China in which the majority of the funding and talent comes from the United States, but the production package includes substantial financial investment from China.[5] These negotiations also involve content, hence indicating a deepening US-China relationship in the realms of film production and consumption.

In another study emphasizing a political economy perspective, we find Chinese government organizations such as the CFGC and the State Administration for Press, Publication, Radio, Film and Television (SAPPRFT) often have final cut on these productions.[6] This trend allowing Chinese authorities control over coproductions extends to explicitly censoring content to suggesting appropriate strategies to represent the country through coproductions, providing for a mechanism whereby Hollywood companies have to abide by the regulations of Chinese authorities.[7] Consequently, these initiatives led to changing modes of representation and, by extension, potentially transformed public perceptions of China.

The influence of these cinematic arrangements is, understandably, a potentially contentious matter with considerable political and reputational implications, especially for producers in Hollywood. The films produced in this context often take on activist political inclinations where they comment on notions such as democracy and individual freedom, sometimes in very obvious confrontational ways, at others on more subtextual levels.[8] These films want to be relevant, and thus it is not uncommon for them to take on global crisis narratives and political flashpoints.

Environmental crises, particularly climate change, are arguably the most imminent and urgent of these themes. Films like *The Day After Tomorrow* and *An*

Inconvenient Truth are just some of the better-known examples of films confronting these themes. Similarly, the conflation of cinema and environmental criticism has resulted in considerable academic work in the field of *ecocinema*.[9] While these studies tend to focus on analyzing the content of Hollywood films as reflections of ideological norms, they seldom focus on the production histories or policy contexts of films. This chapter addresses this gap by suggesting that such contextual factors matter for the directions environmentalist arguments take, and that the environment appears to provide an apolitical meeting place on which to base these coproductions.

Furthermore, filming locations are a key negotiation point of coproductions, and thus, it perhaps isn't a coincidence that the environment arises as a key narrative trope in many of these coproduced films. As I will show, the environment becomes a critical trope of coproduction narratives, often reflecting crisis not only as a backdrop but also as vital to the plot. How locations are displayed and discussed in relation to crisis are telling of cultural negotiations about the past and the future, declining and rising empires, and the ways cinema both reflects and shapes audience expectations. As Fredric Jameson has suggested, science fiction as a narrative genre is particularly well suited for criticism of the present-day world as it can be used to displace explicit political criticism into the future in ways that avoid direct politicized confrontation in the present day.[10] Thus, science fiction as a genre provides a productive field to navigate coproduction politics and cinematic desires insofar as it enables films to address natural resource crises, yet to do so in less confrontational ways, as the plots are written in the future and do not explicitly engage contemporary political flashpoints.

From Production to Content Management

Coproductions between the United States and China's film industries have increased as the rise of Chinese film audiences in the first decade of the twenty-first century have grown from an insignificant percentage of the global market to challenging the economic and cultural centrality of the US box office. In the early 2000s, the revenues from the Chinese box office grew at an annual rate of 30 percent or more, reaching 2 billion USD by 2010. While *Looper* did not ultimately attain full coproduction status, the international partnership allowed it to be distributed in China without any of the usual restrictions facing American film productions.

The growth has only escalated since *Looper* opened in 2012 as the total market has grown to 10 billion USD by 2017.[11] Because the United States and China are already the two biggest cinema markets in the world, such collaborations are worthy of our attention—the future is literally being imagined in these two massive film markets.

These collaborations have in turn increased trade negotiations between the countries to allow more American films to enter the Chinese filmgoing market, which continues to rely on protectionist regulations to safeguard the domestic film industry from competition with imported products. Negotiations have been conducted through the World Trade Organization (WTO) and other bilateral communication at the highest levels of political power, including negotiations between US Vice President Joe Biden and Chinese President Xi Jinping.[12] Currently, the total quota is at thirty-four imported productions per year, a figure that might seem unsatisfactory from an American perspective, as this creates a considerable restriction on the box-office potential of the hundreds of major films produced by the industry on an annual basis.

Accordingly, increased negotiations and the development of diverse tactics for market entry have been devised. Coproduction ventures between companies such as Legendary Pictures and Dalian Wanda International (DWG), as well as an increased exchange of talent between the industries are some of the indications of these dynamics of increased competition and collaboration between the United States and Chinese film industries. Hollywood's struggles to secure a share of the exponentially increasing Chinese box office has translated to explicit content and production management strategies to both avoid these impositions of protectionist policies and target the newly forming taste of Chinese film audiences.

In discussing the recent emergence of intense film coproduction and collaboration activity between the United States and China, Kokas suggests that they closely align with the PRC's financial, technological, and policy priorities.[13] These policy movements have to be reflected in film content, as alongside economic competition, ideological and other political factors in film narratives tend to cause the most amount of friction between the parties. Accordingly, as may be obvious from the above examples, the genre of these Sino-US coproductions tends to invariably be science fiction. Not only does this genre sell particularly well with Chinese audiences, it also provides a plausible landscape for fantastical narratives involving time travel and giant robots battling interdimensional monsters.

Indeed, science fiction has been a prominent theme in ecocritical literature and film studies, which is not surprising considering how its speculative and imaginative qualities enable creatives to address concerns that in many key ways reflect contemporary societal problems covering areas from the climate to political turmoil.[14] Setting events in the future allows filmmakers to avoid potentially problematic aspects of contemporary society. Certainly, placing these narratives in the realm of science fiction allows the productions to avoid politically sensitive areas but also to touch on themes relevant to contemporary social and cultural concerns, such as climate change and the role of clean technologies in complementing an increased awareness of environmental regulations, as well as allowing the films to include and exclude (often stereotypical) elements of Chinese culture in ways that do not appear obtuse or overbearing for the films' audiences in both the United States and China.

At the same time, the use of futuristic themes paints China as a high-tech landscape investing in and benefiting from innovative approaches to societal and technological development. This angle provides a particularly potent example of how the politics of coproductions are inherently about a pragmatic promotion of various stakeholder interests and notions of competitiveness, but these areas look different in their integration into the films' diegesis. The translation of ideological and commercial interests into film content requires elaboration on their potential political significance for Chinese, US, and global imaginaries and for emergent ways to analyze these political affiliations.

These concerns are particularly intriguing when considering how these narratives highlight STEM thinking. STEM is a term used to define developmental and environmental priorities in public policy. In the humanities, the concept is often derided for its explicitly pragmatic orientation on quantifiable measurements and enthusiastic prioritization of more economic outcomes.[15] While STEM strategies frequently reflect the realities of social and economic policy, the arts and the humanities sector has proposed a variation on the concept: STEAM (Sciences, Technology, Engineering, Arts, and Mathematics).[16] The idea is to integrate some of the more critical analytical perspectives of the humanities into thinking about the contributions of STEM topics. In an intriguing twist, this approach is often replicated in the films under interrogation. For example, a production like *Looper* displays an enthusiastic reverence for themes concerning scientific and technological innovation while grounding them in dramatic human narratives. Of course, this does not make the film a part of the STEAM "agenda," but rather, provides a reflection of the general political and cultural circumstances that facilitate the coproduction of films between the United States and China.

Global Ecocritical Perspectives

While setting these narratives in China makes sense on the basis of audience demographics alone, I want to suggest a further rationale for these practices based on a direct correlation between film content and the establishment of policies related to coproduction. The US-Sino coproductions act as intriguing reflections of STEAM politics because they raise questions about political and scientific responses to global warming, climate change, and the state of humanity as part of a fraught ecological framework that is reliant on, precisely, human intervention to undo some of the problems caused by humanity's developmental overreach. By addressing these developments through a cultural forum, these films, whether they intend to do so or not, problematize and thus thematize debates revolving around human-made climate change, and in doing so, they bring together cultural and political value systems from the two key players in these negotiations, the United States and China.

To evaluate these depictions in a wider sociopolitical framework, I turn to the field of ecocinema studies, which aims to understand the complex roles of humanity in the ecosystem from a multitude of angles.[17] These range from films dealing with environmentalist themes such as *An Inconvenient Truth* or *The Day After Tomorrow* to films capturing lifestyle choices and consumption habits.[18] According to Rust and Monani, more or less all films can be considered in an ecological framework as they inevitably reveal significant aspects of the ways humanity understands its role in the ecosystem.[19] Films tell us about attitudes to lifestyles and to pervasive and dominant value systems, and by doing so they provide a valuable field of analysis to understand the ways cultural inflections shape societal understandings of key themes related to environmental matters. For example, we may turn to films with no particular environmental content, such as the *Fast and the Furious* franchise, to understand the ways popular culture perpetuates hyperconsumptive practices, or we can discuss the military industrial complex and resource politics via analysis of the far-flung galaxies portrayed in fantasies like *Avatar.*[20]

The point is that ecocinema provides a vivid and far-reaching venue to explore the complex mechanisms through which the resource-intensive role of humanity is explained in more or less explicit ways to itself. These concerns include the consolidation of environmentally harmful uses of natural resources, such as fracking, or the failure of intergovernmental climate negotiations to overcome petty political differences, as seen in documentaries from *Gasland* to *An Inconvenient*

Sequel: Truth to Power.[21] Similarly, environmental politics in China have received coverage that explores the exploitative role of Western companies operating in China (*The Red Forest Hotel*) to debating China's deadly air pollution (*Under the Dome*).[22] These films thus touch on contemporary environmental politics in ways that reflect an ecocritical perspective on contemporary social and environmental problems. Yet, how do these ecocritical concerns connect with our focus on US and Chinese coproductions and the transformations in China's international image? How do the thematic patterns they display rely on grounding their politics in an ecocritical framework?

To begin to answer these questions, I now turn to two ecocritical themes prevalent in these coproduced films: Chinese scientists and the environmental conditions they inhabit. While only two of a range of tropes we could focus on, they illustrate many of the cultural biases and norms that are common in these coproduced films. Both are also revealing about US assumptions and attitudes to STEM thinking and, I argue, indicate how these US-China coproductions mobilize STEM aspirations to portray miraculous scientific advancements that also minimize political friction, thus facilitating access to Chinese production support and Chinese markets.

Chinese Scientists on Hollywood Screens

Since coproduction arrangements between the United States and China became more standardized, power dynamics between the films' protagonists have become more even. A key strategy here is not relegating Chinese characters to only communicating in English, but now allowing them substantial dialogue in Mandarin. An example of this strategy from comparatively early on is *Iron Man 3*, a part of the Marvel superhero franchise, where genius and technological mastermind Tony Stark (Robert Downey Jr.) of the intergalactic band of heroes The Avengers journeys to Beijing to consult with Dr. Wu (Wang Xueqi) about overcoming his foe, The Mandarin, and removing shrapnel from his chest. The only place to perform this dangerous operation is with Wu in China. Their partnership is depicted in a scene that is brief in the conventional international version, but that is extended to a four-minute scene in a bespoke Chinese version containing Stark touring the facilities and meeting Wu and his various associates. The scenes were not particularly well received in China and were often criticized for not contributing anything to the narrative, especially as the scenes contain egregious product placement with Wu and his staff consuming Yili milk.[23] Yet, from the perspective of Marvel Studios, inserting the scene achieved what it set out to do: enabling a photo shoot of Iron Man outside the Forbidden City and including a substantial scene catering only for Chinese audiences, thus establishing a direct channel with this all-important demographic.

Linguistic address at Chinese audiences is done for cultural reasons, at least according to the China Film Co-Production Corporation (CFCC): "Undoubtedly, the demand for Chinese content is growing, and the best way to meet that demand is to directly access China—the only authentic source for Chinese culture, creative and expert resources."[24] Whether this is ultimately a form of cultural exchange diversifying the representational scope of Marvel's then-contemporary limited focus on white ethnic characters, or just a market gimmick for a major studio to connect with an emerging audience base by providing largely stereotyped impressions of Chinese culture is up for debate. Yet, the *Iron Man 3* scene is significant as it provides a historical example of a developing production incentive where the image of China, as was also the case with *Skyfall*, is distinctly positive and progressive in tone, creating impressions of a country defined by megacities and advanced sciences. Wu, as both a conscientious and innovative scientist and mentor figure, challenges the geopolitical imaginary implicit in the Iron Man comic narratives, where the villain The Mandarin was depicted as an Asian man (transformed here into British actor Ben Kingsley, performing an ethnically undefined role under the command of the film's real villainous mastermind, American scientist Aldrich Killian, played by Australian Guy Pierce). This could be a case of political correctness, but Wu's prominent role in the Chinese version tells a different story—one where the ethical impetus of scientific advance is shifting from the United States to China. This leads us to a significant theme of this chapter: how Chinese sciences and technology are portrayed in these coproductions. Much as the megacity spectacle of these films' visuals replace the "touristic" scenery of the Orientalist imagination prevalent in many of the antecedent depictions of China, the focus on Chinese protagonists as leading scientists and experts proposes a fundamental change from the kung fu masters or dubious politicians seen in a range of Hollywood films from *The Manchurian Candidate* to *The Chairman*.[25]

These ideas prevail in films from *Looper* to *Transformers 2*, all coproductions with distinctive input from Chinese financiers but with creative input dictated overwhelmingly by above-the-line US talent. Similarly, *Transformers 4: The Age of Extinction* had most of its latter half set in Hong Kong, a consequence of a coproduction partnership between Paramount Pictures and Jiaflix, resulting in yet another depiction of China as an advanced society using the most up-to-date technology and research and development, especially in relation to natural resources and energy.[26] In a typical science-fiction dressing, KSI industries are harvesting a substance called Transformium and using it to build artificial transformers in a Hong Kong high-tech factory headed by lead scientist Su Yueming (played by Li Bingbing). In comparison, lead protagonist Cady Jaeger (Mark Wahlberg) lives in a rural farmhouse, and substantial parts of the film take place in US Midwest settings (in marked comparison to the rest of the franchise entries). The contrast between the United States (flat rural cornfields and empty highways) and Hong Kong (tall skyscrapers and advanced technology) could not be more pointed—hence indicating that even in these STEM-inflected and international coproductions, the films rely upon the worst kinds of national stereotypes and cultural misperceptions.

In the case of Hong Kong, the obvious justification for its glamorous portrayal (despite the fact that it is among the least egalitarian cities in the world) is that such imagery enables a US product to overcome problems with access to the Chinese markets. One could also suggest that there may be some level of understanding of China's advances in investing in science and technology as well as developing environmental regulation and management protocols to meet the pressing needs of its population; yet this too, as discussed elsewhere in this collection, bears little resemblance to reality, as Hong Kong is a well-documented environmental disaster. The rising immediacy of the climate crisis in the Hollywood imaginary has been mainstream material at least since films like *The Day After Tomorrow* became a box-office sensation; but as regulatory advances through the United Nations and the Intergovernmental Panel on Climate Change have struggled to establish functioning protocols, the role of both Chinese private capital and state investment to address these unique challenges has increased, especially in terms of technological advance, which is where many of the key battles will be waged. The effects of climate change have emerged as visible parts of daily life around the world, but these have been especially pronounced in China, where, for example, as many of the chapters herein demonstrate, air pollution and toxification of waterways have consolidated into hotbeds of public discontent, leading to the necessity of establishing viable strategies to combat them.

Here, it is worth noting that much of the US media industry reflects specific values at play in liberal Hollywood, including the role of climate change in global politics. Whereas American society is seen as lagging in its ability to respond to these evolving challenges, the China of these films operates right at the forefront of new technology. Yet, while the industry seems to look to the East to provide new visions for societal and technological progress, we need to be careful of essentializing the complex impulses of the industry under a singular, "liberal" label. For example, these films continue to treat the military industrial complex in distinctly glamorized ways in films like *Transformers 2* and *4*, but at the same time, the historical transition in the Chinese imaginary they exhibit does reflect a more advanced perspective on China, one that is modified to meet the country's economic importance. Thus, I suggest that the turn to China in Hollywood's production and representational tendencies plays out these politics in intriguing but also complex, and not necessarily consistently environmentalist, ways.

These developments can be traced especially in the evolving treatment of China in the *Pacific Rim* franchise. Produced by the Legendary film studio, the first film of the franchise taps into imagery and thematology familiar from the *Godzilla* and *Transformers* franchises to produce a blockbuster event predominantly aimed at the Asiatic market. The narrative emphasizes international collaboration, with Americans joining forces with a multinational crew of pilots for the Jaeger robots, a weapon large enough to combat the interdimensional monsters attacking the Asiatic region.[27] The film emphasizes military might, leading some in China to criticize it as propaganda (see Kokas), and its view of China also falls into largely superficial spectacle.[28] Significantly for our purposes, such considerations are directly addressed in the film's sequel, *Pacific Rim: Uprising.*[29] The majority of the film takes place in China (mirroring the acquisition of its production company Legendary by DWG) and other parts of Asia. While American and British soldiers remain leads, the Chinese scientist Zhao (Jing Tian) has developed new technology that allows the pilots to harness their war machines much more efficiently. The military predictably has questions about implementing a new direction for their industrial program, and as sabotage hampers the machines, suspicion turns to Zhao. A twist soon confirms that a leading member of the American scientist team is collaborating with the enemy, suggesting, much in contrast to the first film's fetishization of military might, that the aggressive shock-and-awe approach to battling this universal foe is insufficient to control the technology or coordinate the efforts against the monsters.

Through this turn, the depiction of China becomes much more complex as it not only provides a passive pictorial backdrop (a gesture in the first film that amounts to a tip of the hat for the financial support for the production) but also transforms to an active participant when Zhao is revealed to be a conscientious and ethical businesswoman who has developed advanced, energy-intensive tech, but who also faces substantial skepticism from American military leaders, who cling to outdated strategies. Zhao's plan facilitates the right kind of combination of her advanced technology and human ingenuity to defeat the supernatural foe. Business and engineering ingenuity combine as a feasible alternative to the simplistic use of force, enabling these films to model plausible answers to existence-level catastrophes. Even while actual governments squabble, the film depicts a multilateral mode of collective problem-solving via goodwill collaboration. The *Pacific Rim* franchise thus showcases a transition in representational attitudes, where Chinese contributions to global goals are at the forefront of innovation.

Displaying Chinese Soft Power

While highlighting the ways the Chinese private sector leapfrogs over developmental goals, these coproductions also frame the actions of Chinese governmental officials and departments in beneficial ways. This enables us to evaluate a range of coproduction incentives with marginal environmentalist content as part of this lineage of ventures mobilizing ecocritical content to patch over potential political differences. The 2015 blockbuster science-fiction film *The Martian* foregrounds these ideas in intriguing ways.[30] The setup of the film is ideal for an ecocritical reading and to illustrate my thesis on how these coproductions use natural resources and technology development as a means to consolidate collaborative approaches between the United States and China. While neither *The Martian* nor *Pacific Rim: Uprising* concern themselves with explicit environmentalism per se, they fit well with the earlier arguments about all films revealing aspects of human approaches to environmental resources, as they display dominant attitudes about value systems or behavioral norms, or indeed, about perspectives on beneficial scientific and technological development and collaboration.

Accordingly, the narrative of *The Martian* focuses on a multinational mission to explore the surface of Mars. The crew is made up of a diverse range of Western scientists and military astronauts (and is notably lacking any Asian participants) and arguably is aimed at showcasing contemporaneous impressions of the International Space Station crew. The mission is cut short by a planetary storm as the American astronaut Mark Watney (Matt Damon) is stranded alone on the planet and has to survive on only meager resources and with relentless ingenuity. This requires an extensive amount of innovation on his part, including the now famous line—used in schools everywhere—that the desperate astronaut will need to "science the shit out of this" if he wants to survive, for he must figure out how to geoturf a planet where nothing grows while rationing air, water, and food. The survivalist narrative evokes a pronounced sense of the frontier spirit, with Watney's ingenuity enabling him to save his own life and reach a point where he can be rescued by an international mission (again visibly and emphatically composed of distinctly Western personnel).

Yet, American ingenuity gets him only so far, as the propulsion engine required for the rescue mission fails its endurance test and the whole venture is jeopardized. At this key point, Chinese ingenuity comes to the rescue of this all-important venture. The film cuts from the NASA headquarters and crammed science labs to an external shot of the CNSA (Chinese National Space Agency). The visuals often hark back to rather conservative depictions of Chinese society, doused in monochromatic blues covering monolithic, brutalist architecture. Yet, inside we are given a glimpse of an advanced technological setting, led from a glistening glass office by the department head in charge of the space operations who comes up with the idea of contacting NASA to offer them the use of their space technology as a form of cooperation between the two agencies. His aide asks whether these might contradict national security regulations, but in a truly altruistic turn, the Chinese head decides that scientific exploration and humanist value systems are more important than interstate bureaucracy. Subsequently, the NASA heads journey to China, where they monitor the final attempt at the rescue mission. The film touches on small cultural differences between the representatives of these two national agencies to provide humor to the narrative, but overall the film offers a serious, complementary approach to international cooperation through advanced technology. In terms of US-China relations, then, *The Martian* offers a hopeful view of collaboration for the common good, with STEM expertise and good old-fashioned grit combining to turn a tragedy into a triumph.

While the film and the other examples addressed to date do not provide orthodox climate-fiction narratives, and while they often feature only incidental material on environmental concerns, they do end up conveying attitudes and perspectives that favor collaborative ventures in technological advance. Furthermore, the way these collaborations are foregrounded as considering panhuman concerns allows them to contribute to a form of mediated cultural diplomacy, an ethos of collaboration where access to Chinese markets transforms Hollywood content and plays up the ways these ventures consolidate Chinese soft power. We get a sense not only of fundamental transformations in imagining China, but also of shifting Hollywood perceptions of Chinese private and state operations. The emphasis on educational advances provides these films the chance to display their coproduction partners in a positive light, but also to tap into general public perceptions of China's advances in math and sciences. These depictions are obviously both economically and culturally sensible, yet when we outline this history, from the early examples of *Transformers 2* and *Skyfall*, focusing simply on depicting China as an exotic setting, to the more recent examples of *Pacific Rim 2* and *The Martian*, where Chinese innovation is celebrated, we get an intriguing challenge to the ways cinematic ecocriticism has been understood so far by academics. If Hollywood films have highlighted US political and cultural activities as key instigators of awareness and debate on environmental issues (as some, including Taylor and Brereton, have suggested), these coproductions show us an impression of international collaboration where the United States is distinctly lagging behind.[31] Even as they emphasize the necessity of collaboration on environmental issues, they act as a form of environmental advocacy and soft-power negotiation providing China with a more pronounced role on the global stage. And while the economic necessities of a film industry in transition may be driving these changes—which are still, of course, produced largely by American creative talent—it is worth asking if such transformations also make diversifying not only the portfolio of film texts analyzed by academics but also the policy infrastructures facilitating the production of films much more of a pressing concern.

A New World is Possible

Following these wider patterns of development, I now turn to two case studies reflecting these transformations. The films *Skyscraper* and *The Meg* are recent coproductions featuring major international stars Dwayne Johnson and Jason Statham, respectively, and focus on stereotypical blockbuster narratives, albeit now repositioned for the emerging Chinese audiences. This section will especially focus on exploring how encounters between American and Chinese protagonists and contexts challenge much of the existing thinking on US global leadership, a notion potentially reflecting the need for these coproductions to appear politically neutral in preparation for state censorship, but also as a powerful indication of the ways leadership in technological innovation paves the way for a shift in global power dynamics. Drawing on themes raised thus far, I will consider how the plots focus on displaying China as an advanced technological civilization, already prepared for the incoming green energy revolution, one that is developing solutions to pressing global problems such as overpopulation and mismanagement of natural reserves. At the same time, it is worth maintaining a critical perspective on these representations, especially as the role of environmental concerns in both films tends to be marginalized by their adherence to blockbuster narratives.

Skyscraper focuses on former counterterrorist officer Will Sawyer (Dwayne Johnson), who, after a botched military operation, is provided the opportunity to take charge of security at the tallest building in the world, The Pearl, located in Hong Kong.[32] In *The Meg*, the protagonist is Jonas Taylor (Jason Statham), a marine operator who burns out following a deep-sea diving mission during which a nuclear submarine is attacked by a humongous sea creature. Both protagonists are masculine caricatures of Hollywood action heroes who have to take jobs in South Asiatic locations as a result of crises in their previous work environments, thus giving in to stereotypical perceptions that jobs in these geographical environments are something of a downgrade from their crème-de-la-crème industries in the West. Undoing most of these stereotypes about locations for advanced industrial labor, both films show these washed-up Western professionals now finding work in high-tech China. They have become used to "doing things their way," often through distinctly down-to-earth (read manual or crude) methods. In many ways, they embody the common tropes of the action film genre's focus on working-class, hard-bodied heroes whose hypermasculine presence acts as a focal point of narrative motivation as well as providing an identifiable set of marketing strategies.[33] Yet, when they enter these new labor conditions, they have to adapt to working with advanced technology and complex cultural politics—creating a new crisis for both protagonists that requires flexibility and resilience.

Crisis, as we know, requires intervention, and thus, both films highlight the ways these "dinosaurs" try to acclimate to their much more dynamic new surroundings. But whereas their previous lives had revolved around personal crises, they now encounter crises of an existential kind, even if these only appear as a side theme of the respective films. In *Skyscraper*, for example, The Pearl is a state-of-the-art architectural achievement, featuring, amongst its glistening, spectacular exteriors, a full-scale botanic garden that cleans the air inside the building. It is the brainchild of billionaire property developer and engineer Zhao Long Ji and generates all of its electricity through "a double-helix wind turbine," as we are told by an in-film advertisement for the building. Ji's construction is designed to overcome the environmental problems facing contemporary China and Hong Kong, specifically overpopulation and pollution. Yet, it is also clearly a product of the unequal balance of affluence in these societies, where the upper echelons of society are able to consume the purified air hundreds of stories above street level. While *Skyscraper* does not quite get to a point where it would be explicitly critiquing contemporary social inequality, it does thematize the city of Hong Kong as an existential crisis zone, where overpopulation and economic inequality feed off one another.

Likewise, in *The Meg*, Jonas is brought back into "the game" due to his previous encounter with the prehistoric creature. The film takes great care in introducing us to an underwater base exploring a previously undiscovered crevasse in the ocean through Jonas's no-nonsense perspective. The base is another glistening setting that showcases the most advanced research environment for marine biology, which the film tells us has been designed by the Chinese engineering talents Dr. Minway Zhang (Winston Chao) and Suyin (Li Bingbing), whose guidance and management of both technology and staff enable the laboratory to meet ethical guidelines for underwater exploration. The base is portrayed as free from explicit commercial pressures, yet it is also populated by an odd American billionaire who appears to want to harvest the science for commercial purposes—in this way, American capitalism lingers in the background as a threat to scientific inquiry. This scientific exploration for new ecosystemic worlds and energy resources turns into a battle for survival as the scientists breach a previously secluded undersea environment, unleashing a giant megalodon, a pre-Jurassic shark, that emerges from the crevasse to wreak typical blockbuster havoc.

These films are particularly productive illustrations for our purposes as they provide recent iterations of established Hollywood genre templates transplanted to China, and where the crisis narratives are framed through an environmental lens. *Skyscraper* is an obvious update of *Die Hard*-esque action films with a lone hero facing off a terrorist gang, whereas *The Meg* draws inspiration from horror films such as *Jaws* and *Deep Blue Sea*.[34] The depictions in both films adhere to the principles we've identified in these coproduction strategies. In both films, China is shown as a space harnessing advanced technology where benefactor private investors push science in new directions to meet contemporary needs. In contrast, the United States is depicted in *Skyscraper* as a fraught domestic space dictated by socioeconomic trauma, and, in *The Meg*, as jeopardized by morally compromised corporate greed. Consider, for example, the harrowing opening moments of *Skyscraper* in which Will faces off with a recently unemployed father who's been laid off as result of the automatization of the manufacturing sector and by societal infrastructures that lack protective safeguards for ordinary families. Likewise, a morally dubious billionaire, the Elon Musk–like investor Morris (Rainn Wilson), is featured in *The Meg*, initially appearing to support scientific exploration, but who turns out to be a Zuckerbergian micromanager of all aspects of the operations, down to exploiting the now-unleashed force of nature for commercial purposes. Morris plans to film himself destroying the shark by carpet-bombing the ocean with depth charges, which clashes with the intentions of Zhang, Suyin, and Jonas, who try to capture the shark by sedating it, which would be considerably less harmful to the environment than the mass carnage of the depth charges.

While the typical genre structures remain in place for both films, these existing formulas are used to replace established American cultural imagery—such as the glistening skyscrapers of *Die Hard* or the sunny, small-town, beachfront idylls in *Jaws*—with Chinese landscapes and iconography. In *Skyscraper*, a group of international terrorists take over the building to extort money from Ji, who has refused syndicate blackmail in the construction of the building. It is up to Will to fight them off and save his family as well as the reputation of the Chinese entrepreneur, who joins Will in the fight. Thus, Will has to jeopardize his nuclear family unit (pun intended) as he navigates the complexities of The Pearl's innovative energy-generation infrastructure, resulting in scenes that provide the necessary high-tech thrills for a blockbuster action-adventure film, as Will has to navigate the timing of the blades that power the turbines, and that also emphasize the complexity of technological innovation that has gone into constructing these new benchmarks for resource-capture technology.

Meanwhile, *The Meg* uses a range of established genre scenarios from deep-sea submarine attacks to a beach massacre that mobilizes imagery from *Jaws*, Steven Spielberg's seminal blockbuster. Whereas in *Jaws* the beach massacre takes place on the 4th of July, here *The Meg* moves the beach to Southern China, with Chinese revelers (in some cases following expected tropes seen in other films, such as US pop-culture phenomena like *Crazy Rich Asians*) eaten in large numbers by the shark.[35] Unlike *Skyscraper*, *The Meg* makes conscious use of centrally placed Chinese flags to reinterpret the taken-for-granted cultural customs of Hollywood cinema, as well as including substantial dialogue in Mandarin and Cantonese, something that was largely absent in *Skyscraper*. The fact that dialogue is presented in the original language and not morphed into English is significant, as it indicates a conscious acknowledgment of the shift in audiences from the United States to China.

The shifting balances of power in these films, especially as they are so reliant on established genre narratives, situate them as transitional texts. Both were global box-office hits, with *The Meg* accumulating $527 million globally (US $143 vs. China $153) and *Skyscraper* $304 million in total ($67 US vs. $98 China).[36] The totals emphasize the relevance of the Chinese market for these coproductions. Thus, it is not surprising that the films literally enact the transition of global power from the West to the East in a showcase of soft power on a global scale. Showcases of military power take place as the Hong Kong armed forces operate at top standards in *Skyscraper*, and *The Meg*'s Jonas calls in two destroyers from the Chinese military to curtail the problem. The fact that these are conducted within the confines of an American creative industry is revealing, as narrative salvation is gained largely by a combination of American heroism and sensible forms of strategic Chinese collectivism. *The Meg* even ends with an implied cross-national and cross-racial relationship, suggesting—as does *Crazy Rich Asians*—that in the new world of coproduced blockbusters, viewers of all ethnicities, classes, and nationalities can unite around the stereotypical worship of summer-worthy hard bodies.

As such, these films implicitly suggest that we might want to take *Looper*'s advice and learn Mandarin to prepare for a shift in global power relations, a notion that these coproductions reflect both on the level of the text and in their production context. At the same time, the lack of active Chinese creative agency in these representations must be acknowledged. While certainly, the Chinese state will have a very closely monitored say in what is admissible on the representational level, the key creative labor—including directing and scriptwriting—is exclusively done by US-based talent. Thus, films like *Skyscraper* and *The Meg* may tell us more about American cultural obsessions and value systems than anything in particular about how China sees itself. Here, the omnipresence of Chinese capital and governance alters the balance, most often not as a dictated, top-down imposition by the Chinese state, but as a largely voluntary thematic shift on behalf of the US producers positioning products for Chinese audience tastes. This targeting of a significant audience demographic is especially visible in a production like *Skyscraper*, as its roots in Canadian coproduction arrangements do not seem to matter that much (they are effectively erased on the textual level). In contrast, the necessity of emphasizing

Hong Kong as a space of both advanced technological development and effective societal governance emerges as a thematically significant area for the film. Here, the location itself becomes thematized, contributing a potent sociopolitical message, but most importantly, providing a tangible attraction for its target consumers—the Chinese audience.

At the same time, by injecting key advances from STEM thinking into these well-worn genre formulas, both films can be interpreted as offering solutions to the ongoing environmental crisis, even as they use this material as politically neutral ground for audience engagement. While focusing discussion on the environment provides a universal field of concern that can transcend national interests, a much more critical perspective on these tactics is also required, for by using the environment as a safe zone for political games between the Chinese and US film industries, actual environmentalist concerns are sidelined. For example, the notion of The Pearl as a solution to overpopulation is only alluded to in *Skyscraper*, and the building itself seems to be largely inaccessible for ordinary residents, being entirely devoted to well-off clientele. Similar concerns can also be raised in relation to *The Meg*, for while Jonas and the scientific team act with ethical concern for the environment, the film, as an example of blockbuster spectacle, relishes destroying nonhuman life forms and draws most of its genre thrills from the menace untamed nature poses for humankind. Especially in this film, it seems safe to conclude that consciousness about environmental issues is not so much raised as battered. The shallow spectacle of both films may *appear* environmental, but the commodification of environmental themes into commercial spectacle ultimately acts as a concession to the political economy of these coproductions.

While such themes may provide an agreeable means to neutralize political discontent in accessing lucrative audience demographics, what both films seem to forget in all this is actual environmentalist concerns. Thus, the result is environmental content that goes nowhere—hence the use of new technology in *Skyscraper*, for one, not playing any significant role besides as a setting for an action scene, or *The Meg* with its potential thematic content focusing on new energy resources completely sidelined in favor of awe at Chinese technological ingenuity and depictions of the standardization of global elite lifestyles. Even as any concrete arguments over the societal implications of such strategies would have to be examined through audience and reception studies, adopting such an ecocritical perspective suggests the contribution these films make to environmental politics or to advancing STEM education and thinking appears less than the sum of their parts.

Conclusion

Cinema is a vitally important medium for environmental communication to engage a range of publics. Global in reach and massively popular in scope, film affords us the opportunity to consider which crises or fears and which fantasies or hopes are being produced, promoted, circulated, and consumed, and how they link with specific cultural contexts. Ecocritical analysis of films provides a means to investigate these practices, especially to facilitate more comprehensive understanding of the ways contemporary cultural production takes part in some of the most fundamental debates of our times. While they may often appear to only play host to simplistic entertainment priorities in subservience to the capitalist logic of the cultural industries, they act as key motivators shaping popular understanding of themes they represent, and as significant barometers for critical analysts to understand some of the ideological discourses circulating in popular conversation. From films with explicit environmentalist content to ones with no obvious link with environmental issues, the study of ecocinema provides an approach to understanding the roles of the environment and natural resources in anthropogenic and anthropocentric imaginations.

As suggested, coproductions provide means for both hard infrastructure (as in the case of the production studio Oriental DreamWorks) and soft infrastructure (as with the requirements for Chinese talent to be centrally featured in Sino-US film coproductions).[37] The reasons for these collaborative projects appear largely economic, often for understandable reasons. Hollywood wants to be able to tap into the Chinese audience base, which is overtaking the significance of domestic US audiences globally. In order to do so, these coproductions have to abide with complex challenges that arise from tight regulations on importing cultural products to the Chinese market, the specific political requirements for content, different labor laws and rules, regulations impacting the management of foreign companies operating in China, and audience tastes that are no longer content with an exoticized Orientalist view of China.

Thus, it would not be difficult to argue that these coproduction priorities and strategic incentives play a significant role in how Hollywood represents Chinese culture and society. While film scholars suggest that these coproduction arrangements now translate to the Chinese government having final cut on American films, private investment via DWG, for example, has led to more positive evaluations of these arrangements. The idea that we can read the films resulting from these ventures as echoing their particular circumstances of production are reflected by other key players, such as the Huayi Brothers president Wang Zhonglei, who suggests collaboration with his company offers US producers "a better idea of the Chinese market. We will provide them with ideas to produce specialized films for the audience."[38] Collaborative ventures like the Legendary/DWG partnership means Chinese capital flows into these productions, but this understandably provides the Chinese partners considerable oversight of the final product as well as necessitates that Hollywood adapt many of its approaches to depicting the "Other."[39]

Yet, is this a case of American producers simply using an existing repertoire of positive images to navigate through the complex market-entry mechanisms? Or is this an accurate reflection of the transformation in the ways the liberal US media industry now perceives the "East"? A noticeable theme in more or less all of these coproductions is an emphasis on natural resources, technology, STEM thinking, and infrastructure—from *Looper*'s emphasis on Shanghai as a futuristic technopolis to *Transformers: Age of Extinction*'s vision of Hong Kong as a space for advanced manufacturing of natural resources, from *Skyscraper*'s focus on sustainable resources for Hong Kong's overcrowded metropolis to *The Meg*'s vision of the misuse of Chinese technology by "greedy capitalists." Consequently, I have adopted critical tools from the field of ecocinema, which provides an approach to reflect on cinema's—and the national cultural industries of which they are a part—often complex and contradictory approach to natural resources.

While these coproductions can be addressed on the basis of how they reflect cultural diplomacy, approaching them from an ecocritical perspective refocuses our attention on differences in environmental policy between the two nations, with China's advances in establishing protocols to meet climate accords largely superseding those of the United States. These are now reflected in the ways the Hollywood media industry regards Chinese scientific and technological innovation over the apparently dirtier operations of the American protagonists. While there may be more obvious economic reasons for these depictions (such as the lineage of explicitly working-class, down-to-earth heroes in American action-film production), they also communicate transformations in value systems and perspectives that are revealing of some of the predominant transformations in global creative industries. Significantly, they indicate a pattern of global power shifting toward non-Western contexts, and they often frame this transition as a consequence of the investments these countries have made in new forms of technological and scientific innovation, hence portraying Asia and STEM as rising together to face the ecological crises of our times.

In their more hopeful iterations, these films show how the combination of American individualism and Chinese ingenuity can combine to face crises, including those linked, if tangentially, to the environment. While it is certainly worth raising questions over whether these productions can ultimately say anything substantial about environmentalist concerns, being as they are explicitly focused on meeting their credentials as popular entertainment cinema, the fact that environmental themes now appear not only as key narrative components, but also as a productive interface to facilitate coproduction politics is of substantial importance. This exploration of popular coproduced films suggests that their reliance on tropes of STEM thinking in relation to ecological crises emphasize a series of classic tensions, particularly via the portrayal of natural resources as key to negotiations of world power. They reflect wider political movements and attempts at generating climate awareness as they provide powerful projections of cultural hopes and fears that resonate with broad audiences, even while reinscribing tired old stereotypes about gender, class, desire, consumption, and nationality. In terms of how they depict both the present and possible futures of US-China relations, then, these films are both enticing and disappointing, for they marshal the environment not so much as a space of care and consideration as of exploitation and drama, yet they also demonstrate that our very survival depends on cross-national collaboration.

NOTES

1. *Looper*, directed by Rian Johnson (USA, China: Endgame Entertainment; Sony Pictures, 2012).

2. Jonathan Papish, "Foreign Films in China: How Does It Work?," *China Film Insider*, March 2, 2017, http://chinafilminsider.com/foreign-films-in-china-how-does-it-work.

3. Deron Overpeck, "Monitored Relations: The US Film Industry, Chinese Film Policy and Soft Power," in *Reconceptualising Film Policies*, ed. Nolwenn Mingant and Cecilia Tirtaine (New York: Routledge, 2018), 25–47.

4. Aynne Kokas, *Hollywood Made in China* (Oakland: University of California Press, 2017), 24.

5. *Transformers*, 2009; *Iron Man 3*, directed by Shane Black (USA: Marvel Studios, 2013); *Pacific Rim*, directed by Guillermo del Toro (USA: Legendary Pictures, 2012); *The Great*

Wall, directed by Zhang Yimou (USA, China: Legendary Pictures, 2017).

6. Kokas, *Hollywood Made in China*.

7. Wendy Su, *China's Encounter with Global Hollywood* (Lexington: University of Kentucky Press, 2016).

8. Betty Kaklamanidou, *The 'Disguised' Political Film in Contemporary Hollywood* (Edinburgh: Edinburgh University Press, 2017); Richard Rushton, *The Politics of Hollywood Film* (Basingstoke, UK: Palgrave Macmillan, 2013).

9. David Ingram, *Green Screen: Environmentalism and Hollywood Cinema* (Exeter, UK: University of Exeter Press, 2004); Robin Murray and Joe Heimann, *Ecology and Contemporary Popular Film* (New York: Routledge, 2008); Pat Brereton, *Hollywood Utopia* (Bristol, UK: Intellect, 2005); Stephen Rust and Salma Monani, *Ecocinema: Theory and Practice* (New York: Routledge, 2012); Pietari Kääpä, *Ecology and Contemporary Nordic Cinema* (London: Bloomsbury, 2014).

10. Fredric Jameson, *Archaeologies of the Future* (New York: Verso, 2005).

11. Kokas, *Hollywood Made in China*.

12. Ibid.

13. Ibid., 19.

14. Ursula Heise, *Sense of Place and Sense of Planet* (New York: Routledge, 2007); Robin Murray and Joe Heimann, *Ecology and Contemporary Popular Film* (New York: Routledge, 2008); Sheldon Lu and Jiayan Mi, *Chinese Ecocinema* (Hong Kong: Hong Kong University Press, 2009); Rust and Monani, *Ecocinema: Theory and Practice*; Pietari Kääpä, *Ecology and Contemporary Nordic Cinema* (London: Bloomsbury, 2014).

15. Zehlia Babaci-Wilhite, *Promoting Language and STEAM in Education as Human Rights* (New York: Springer Publishing, 2019); Arthur J. Stewart et al., *Converting STEM into STEAM Programs* (New York: Springer, 2019).

16. Charles Travis and Armida de la Garza, eds., *The STEAM Revolution* (New York: Springer Publishing, 2018); Xun Ge et al., *Emerging Technologies for STEAM Education* (New York: Springer, 2015).

17. Sean Cubitt, *Ecomedia* (Amsterdam: Rodopi, 2005); Sheldon Lu and Jiayan Mi, *Chinese Ecocinema in an Age of Environmental Challenge* (Hong Kong: Hong Kong University Press, 2009).

18. *An Inconvenient Truth*, directed by Davis Guggenheim (USA: Paramount Classics, 2005); *The Day After Tomorrow*, directed by Roland Emmerich (USA: 20th Century Fox, 2004).

19. Rust and Monani, *Ecocinema: Theory and Practice*.

20. *The Fast and the Furious*, directed by Rob Cohen (USA: Universal Pictures, 2001); *Avatar*, directed by James Cameron (USA: 20th Century Fox, 2009).

21. *Gasland*, directed by Josh Fox (2010, HBO, USA); *An Inconvenient Sequel*, directed by Bonnie Cohen and Jon Shen (USA: Participant Media, 2018).

22. *Under the Dome*, directed by Chai Jing (China: Ming Fan, 2015); *The Red Forest Hotel*, directed by Mika Koskinen (Finland: Luxian Productions, 2012).

23. "Gangtiexia3 yin guanzhong tucao: zhezhong zhongguo tegongban buyao ye ba [*Iron Man 3* leads audiences to Complain: We'd rather not have this special Chinese edition], "*People's Daily*, May 2, 2013.

24. CFCC, "About Co-productions," *China Hollywood Society*, 2018.

25. The Orientalist spectacle is a trend that continues to be perpetrated in the Hollywood imagination of the *Kung Fu Panda* franchise, for example, initially produced by the Hollywood studio DreamWorks Pictures, but taken over by the co-venture subsidiary Oriental DreamWorks for the production of *Kung Fu Panda 3* (Jennifer Yuh Nelson, 2015); *The Manchurian Candidate*, directed by John Frankenheimer (USA: United Artists, 1962); *The Chairman*, directed by J. Lee Thompson (USA: 20th Century Fox, 1972).

26. *Transformers: Age of Extinction*, directed by Michael Bay (USA: Paramount Pictures, 2014).

27. *Godzilla*, directed by Ishiro Honda (Japan: Toho, 1954).

28. Kokas, *Hollywood Made in China.*

29. *Pacific Rim: Uprising*, directed by Steven D. Knight (USA: Legendary Pictures, 2018).

30. *The Martian*, directed by Ridley Scott (USA: 20th Century Fox, 2015).

31. Bron Taylor, *Avatar and Nature Spirituality* (Waterloo, ON: Wilfrid Laurier University Press, 2014); Pat Brereton, *Environmental Film and Ethics* (New York: Routledge, 2017).

32. Burj Khalifa in Dubai is the tallest building in the world. This fictional tower is boasting to be taller. Notably, outside of cinema, "The Pearl" is a building that is a key futuristic feature of Shanghai's sustainability and cityscape.

33. As discussed by authors such as Yvonne Tasker, *Spectacular Bodies: Gender, Genre and the Action Film* (London: BFI, 1993); and Susan Douglas, *The Rise of Enlightened Sexism: How Pop Culture Took Us from Girl Power to Girls Gone Wild* (New York: St. Martin's Griffin, 2010).

34. *Jaws*, directed by Steven Spielberg (USA: Universal Pictures, 1975); *Deep Blue Sea*, directed by Renny Harlin (USA: Warner Bros, 1998).

35. *Crazy Rich Asians*, directed by Jon M. Chu (USA, Singapore: Warner Bros, 2018).

36. https://www.boxofficemojo.com.

37. Kokas, *Hollywood Made in China*, 22.

38. Wang Zhonglei, cited in Hannah Beech, "How China Is Remaking the Global Film Industry," *Time*, January 27, 2017.

39. It may be worth noting that Chinese coproduction initiatives extend far outside these collaborations with the US film industry, leading to, for example, considerable regional collaborations that feature very different power dynamics. See Stephanie DeBoer, *Coproducing Asia: Locating Japanese-Chinese Regional Media* (Minneapolis: University of Minnesota Press, 2014).

PART THREE

On Futurity

Material Cultural Diplomacies of the Anthropocene

An Analysis of the Belt and Road Initiative between China and Oceania

Junyi Lv and G. Thomas Goodnight

Humans are interconnected by our contingencies and unknown planetary futures. In the twenty-first century, the world faces a series of environmental crises, including air pollution, land degradation, and climate change. In this chapter we focus on how the material cultural diplomacies between China and South Pacific Island nations indicate a number of current environmental crises and future aspirations. China's oceanic engagements with South Pacific Islands are shaped largely by the Silk Road Economic Belt and the twenty-first-century Maritime Silk Road, also known as the Belt and Road Initiative or "BRI."[1] China's BRI promises to spread cultural networks, mapping China's work and goods onto the east-west pathways of ancient caravan trade, thus making a breathtaking Cultural Diplomacy (CD) bid. Launched in 2013, BRI has been described as "the most ambitious project in the early 21st century," and as President Xi Jinping's "most ambitious foreign policy."[2] To facilitate BRI, China has established official collaborative relations with 125 countries and 29 international organizations around the world, spanning five continents.[3] Such a project invites us to consider the principles of the ancient arts of cultural diplomacy and to address how BRI points to new possibilities in green communication, both in China and between China and its international interlocutors.

Cultural diplomacy renews and strategically generates conversations among nations. More broadly, cultural diplomacy works across differences to weave the fabric of common relations in a time of global circulation of people, raw materials, services, and goods. At the same time, the successes of industrial and informational diplomatic efforts have generated globally vast environmental byproducts—diseases, weapons, terrorism, plagues, and droughts that transcend national borders—that jeopardize planetary stability. In this sense, CD points to the efforts of scholars, activists, and diplomats to work toward what Phaedra Pezzullo calls an ethic of care, merging green communication with hopes for international peace.

"Anthropocene" is a term that refers to human activities becoming the decisive shapers of global geologic changes.[4] Unfortunately, the United States has begun to falter at climate action, as Pezzullo notes in the introduction to this volume. At the same time, China aspires to take the lead in global climate-change mitigation. President Xi Jinping defines China's global environmental role as a "torchbearer" rather than as a "leader." Within that framework, we aim to show how lowering carbon emissions can benefit China not only environmentally but also diplomatically.[5] Specifically, this chapter focuses on China's challenges for constructing strong relations with three islands dotting the Pacific Ocean. The BRI is an important global initiative in itself, but we believe the precariousness of the Anthropocene can be more fully addressed by engaging Oceania's frontline communities, hence watching how nations on the edge of environmental catastrophe interface with China's BRI.

China has manifested an increasing interest in Oceania and the Pacific in recent years.[6] Under BRI, climate change promises to become a major cooperation point between China and South Pacific Islands. This offers an opportunity for China to contribute to global climate-change mitigation, to lead through building soft power in the national interest, and especially to help Small Island Developing States (SIDS), which are more vulnerable than other developing countries in terms of sea-level rise. Indeed, the SIDS addressed herein face dire threats to their ability to access fresh water, energy resources, arable farmland, living space, and more, amounting to serious challenges to their sovereign existence.[7] Oceania includes the larger island nations of Australia and New Zealand and the scattered island chains that make up Micronesia, Melanesia, and Polynesia.[8] Fourteen of these islands are sovereign, some are states of other nations, and some are colonies.[9] As sovereign yet relatively smaller nations geographically, Fiji, Tuvalu, and the Marshall Islands offer three instructive case studies, for each one challenges hegemonic continental perspectives in a different way. Island-continent relations unfold in what we describe as, respectively, *historia, mythos,* and *plasma.* Paying attention to the local in an age of global crisis permits us to revisit the possibilities and limitations of material cultural diplomacy in the Anthropocene. Thus, we learn from what Oceania provokes us to consider when connecting land and sea. As this region has long been an area of central concern to the United States as well, our analysis offers a case study in how the future of US-China relations may hinge on the development of green communication in Oceania.

Beyond Hard and Soft Diplomacy

Cultural diplomacy (CD) has been put forward as "the third pillar" of diplomacy, standing alongside the more traditional modes of economic and political diplomacy.[10] It is defined as "a series of acts and institutions" exchanging creative "ideas, information, art and other aspects of culture among nations and their peoples."[11] Whereas hard power uses the traditional means of economic and military power to coerce nation-states into compliance, CD uses culture as social capital and soft power to attract collaboration.[12] These acts and institutions increase influence abroad, build national identity, and improve "foreign investment and mutual understanding of shared cultural values."[13] CD requires multidimensional approaches and sincere interest expressed as material investments.[14] If successful over the long term, it can facilitate trust, understanding, and cooperation beyond formal diplomatic ties, especially among parties where diplomatic relations are less established.[15]

When culture is deployed for political purposes, it becomes reduced to harmful or bland, functional instrumentalism.[16] CD, on the other hand, depends on building shared assumptions that cultural values can be appreciated mutually, that friendly relations can be nurtured based on respect and understanding, and that art, education, and language are significant enough to appreciate differences among cultures and to invite new relationships.[17] Historically, the ends of diplomacy have been defined in a wide range: "exchange of views, negotiations, arbitration, conciliation, mediation to adjudication."[18] The goal of diplomacy is not to achieve victory, but to reach for the kinds of mutually reciprocal arrangements that can lead toward agreement; in this way, CD is deeply communicative, focusing less on outcomes than on processes of deliberation and exchange.[19] Possibilities also exist for CD to carry out operations under governmental motivation and intervention.[20]

In an era of global circulation and climate change, we argue that CD needs to make a material turn. Information codes circulate cultural goods through supply chains across the planet, from aboriginal arts, film and television programs, festivals, museums, language, education, tourism, and national airlines, to pop culture and famous public figures.[21] Supply-chain cultures generate artifacts and performances produced by and consumed in different countries, creating the labyrinth of networks often called "globalization." With the rise of Asia's port cities—Singapore, Shanghai, Taipei, Hong Kong, and many more—these globalizing chains of production and consumption have become both deeply rooted in and contested in China and Oceania.[22]

As a rising power in Asia, China has been actively engaged in CD since its modern founding, especially after 1951, when it launched exchange programs—academic, educational, and cultural—with Communist bloc and non-Communist countries.[23] Approximately 75,000 to 100,000 foreigners from at least sixty-three countries, such as Japan and the Soviet Union, visited China at that time.[24] China also made an effort to demonstrate its constructive role in international relations. In 1955, the Five Principles of Peaceful Existence were highlighted in the Bandung Conference by prime minister Zhou Enlai, who also led China's diplomats to showcase friendly and open intentions.[25] During the Cultural Revolution (1966–1976), China suspended these diplomatic relations, but since the economic reforms and diplomatic opening in the 1980s, China has been repositioning itself as a contributor to world peace with "going global" policies and a series of new initiatives.[26] In this phase, China became aware of the difference between domestically facing and internationally facing communication and started to shift emphasis from one-sided propaganda to mutual communication.[27] Nonetheless, gaps exist between China's self-perception and other nations' understandings of China, and between negative impressions of China as a polity and the variety of purposive activities that define its cultures and society.[28] Within this long narrative, BRI stands as a new possibility, evidence of China's newfound role as an emerging global leader.

Moreover, China's BRI has developed during a period when the United States appears to have reached a temporary hiatus in its international leadership. President Trump's bilateral mercantilism now degrades the creative multilateralism that marked US leadership since the end of World War II. In this new era, China stakes its claim to diplomatic power, in part, on identification with the Southern Hemisphere. This goal is hard to fulfill in light of the potentially growing debt obligation of some of its BRI partners in Africa and Asia.[29] Grants of material resources are useful supplements to cultural diplomacy generally, and, we believe, a material turn is absolutely necessary to help support vulnerable Oceania. We temper this call, however, with the caveat that we cannot substitute substantive material commitments and cultural exchanges with the public-relations mechanisms of instrumentalist advertising, branding, public relations, and propaganda. Rather, we call for creative acts of diplomacy, particularly between climate-change vulnerable islands and mainland carbon-generating nations in the twenty-first century.

Climate Vulnerability and Future Opportunities

To some continental residents, the sea remains a measure of distance, either a remote and romanticized paradise or a barren and deep barrier. In old times, China had little motivation to cross the great barrier, the Pacific, and hence interacted little with the South Pacific until the nineteenth century, when Chinese crossed the Pacific to migrate.[30] The South Pacific Islands belonged neither to the land-based nor maritime Silk Road.[31] Even the most substantial maritime activities did not involve navigation to the South Pacific; instead, ships of the previous dynasties steered east to Japan, and south and west to India.[32] Most historians agree that throughout its ancient history, China was neither interested in maritime exploration nor engaged in seafaring power projection. But during the 1970s, China initiated diplomatic relations with the South Pacific. Fiji was the first country to achieve formal diplomatic ties with China. Since then, the communication and interaction between the two regions has become increasingly active. China built formal ties with ten other island countries after Fiji.[33] Those activities were motivated to win international recognition.[34] And so, since the 1970s, China has slowly but surely sought to establish itself in Oceania, both building relationships with individual nation-states and seeking to counter the United States' traditional dominance in the region.

China and the BRI are now moving east diplomatically. The purposes of interaction and communication are not only economic, political, and cultural, but more importantly, environmental and material. Fiji has joined BRI, representing Oceania and the South Pacific Islands.[35] A phrase, "the area naturally along the line of 21st Century Maritime Silk Road," was coined by President Xi Jinping when he first visited New Zealand in 2014, and was then repeated by Chinese economists and think tanks to emphasize the relationship between China and the South Pacific region.[36] As is true with all of President Xi's prognostications, this one must be taken with a dose of skepticism; still, Fiji's prime minister has noted that BRI provides useful opportunities to trade with China and for Fiji to develop.[37] Climate change has been addressed by the BRI. On the website of the Ministry of Commerce of the People's Republic of China, an article titled "South Pacific Area and 21st Century Maritime Silk Road" states that "South Pacific countries are sensitive to climate change. China's progress of 21st Century Maritime Silk Road in South Pacific area should focus on climate-change mitigation, new energy, public health, ocean economy, trade upgrade and investment, tourism, and infrastructure construction, under the regime of South-South cooperation."[38] To pursue such claims, China has become a dialogue partner in the Pacific Islands Forum (PIF); at their 24th meeting, China announced that it would provide RMB 200 million to help mitigate climate change.[39] With the BRI, then, the PRC is making a move toward global leadership in green communication while also seeking to shore up its political relationships in Oceania.

According to the Intergovernmental Panel on Climate Change, global sea levels will increase 58 to 88 centimeters by 2100.[40] Perhaps nowhere are the impacts of climate change more profound than in Oceania. In 2009, the first climate-change refugees were forced to leave Carteret Islands in the South Pacific because of rising sea levels.[41] The water has devoured five Solomon Islands.[42] Other disasters caused by climate change (such as ocean acidification, land degradation, and flooding) threaten local water and food security. People's health, education, and opportunities to live decent lives are at risk.[43] Small island nations are extremely vulnerable. The world consensus is that a 2-degree Celsius (3.6 Fahrenheit) increase above preindustrial levels will be dangerous for many regions, but for small island nations, the threshold is even lower—1.5 degrees Celsius (2.7 Fahrenheit). Global temperatures have already increased 1 degree Celsius since then, which means small island nations only have a half-degree room before disaster.[44] The Alliance of Small Island States (AOSIS) has therefore requested that the final Paris Agreement limit temperature increase "well below" 2 degrees and to "pursue efforts" to strive for 1.5 degrees or less.[45]

Unfortunately, one of the largest carbon emitters around the world, the United States, has since dropped out of the agreement; hence, climate action goals to protect these islands are even harder to achieve, despite the United States having a vested interest in Oceania (including protecting its newest state, Hawaii, and its colony of Guam).[46] Thus, even as China rises toward green communication leadership, the United States shirks its responsibilities. On the other hand, it is important to remember that China has become the world's largest carbon emitter as a result of its rapid economic growth.[47] Environmental crises abound. In 2007, the *New York Times* published a collection of articles called *Choking on Growth.* China is facing problems including air pollution, water contamination, rare species extinction, and waste disposal.[48] Smog envelops major cities such as Beijing and many other regions, threatening health.[49] Coastal China holds the most densely populated regions in this country; the rising sea level directly influences hundreds of millions of lives and their cities. Over one-third of China's carbon emissions result from manufacturing foreign goods.[50] China's decisions, therefore, have impacts well beyond their borders. One of the key questions of the twenty-first century, then, is: Will China's green communication match its actions? Will all the talk about assuming global leadership manifest in sustainable policies and practices?

In response to these questions, the Chinese government launched its China's National Climate Change program in 2007 and established the Ministry of Environment Protection in 2008.[51] These efforts work to mitigate environmental problems, especially climate change. In 2014, China also amended its environmental laws to give "more punitive powers to environmental authorities," and to implement "a broader range of actions for environmental organizations" that define "geographical 'red lines' where the area's ecology requires special protection."[52] These top-down policy-driven actions, however, are "not just economic but ideological." Sometimes, it appears that launching environmental protection programs function more for city branding and marketing than for facilitating genuine change.[53] For instance, during major events such as the Beijing Olympic Games and the Shanghai Expo, air pollution was held in check. The term "APEC blue" was coined to describe the fleeting blue sky in Beijing during APEC meetings in 2014.[54] Sze claims that China's environmental protection relies heavily on engineering, authoritarian structures, and "harmony" discourse. She believes China's eco-cities projects cannot contribute to environmental protection and climate-change mitigation.[55] Like Sze, many observers worry that China's emerging green communication is therefore less about saving the planet than saving the Party.

From this perspective, islands and continents are not only physical masses but also semiotic spaces influencing knowledge of and interpretations about individuals, groups, and populations. To Sloterdijk, islands also have abstract meaning as "miniatures of worlds that can be inhabited as world models." He differentiates three kinds of islands: the absolute island as physical mass inserted into hostile, unlivable environments; the relative islands that serve as cocoons for more fragile exotic life; and the anthropogenic islands, which allow more sophisticated, robust, and adaptive life forms.[56] The categorization reflects his thought of "material places transmuted by an imaginary placenta," including "discourses, forms, and social visions."[57] Within this model, the continental bias must be adjusted to appreciate island vulnerabilities.

For island nations, climate change results in the irreversible loss of homeland, which is not about "economic competitiveness or a global power play," but "life and death." When the water finally rises, the established livelihoods are doomed, and the islanders may have no choice but to leave or perish. The irreparable extinction of an island-chain culture and sovereign identity becomes increasingly tangible as the waters rise, posing a strangely familiar dilemma for environmental communication scholars who study argumentation.[58] Indeed, J. Robert Cox details three aspects to explain the idea of "irreparable."[59] One is *uniqueness*, which speaks to experiences that are often more valued than repeatable ones in our lives. Loss of the unique is particularly painful, especially losing it to the usual, the ordinary, or even the vulgar. Another aspect is *precariousness*. The sense of imminent loss generates anxiety about more limited choices for the future. Being irreparable is about diminished options. The last aspect is *timeliness*. When the choices are constrained, time becomes more urgent. Choosing the right timing to act may be a way to compensate the lost space for choice. In the case of the Pacific Islands, nations and cultures are becoming irreparably threatened by climate inaction. The stakes of climate negotiations, therefore, cannot be more serious for their futures.

To confront these crises, alternative rhetorics are needed; though they may be counterintuitive to some, we now turn to ancient rhetorical terms to help us consider how to map the uncharted future of cultural diplomacy. According to Hellenistic and Roman doctrine, three literary forms are differentiated by how they are situated toward what is, toward what is not, and toward what yet may be: *historia* (fama, verum) identifies what is reported and true; *mythos* (fabula, falsum) stretches into the realms of fantasy and tales; and *plasma* (fictum, argumentum) suggests through the likely or fictive that which may be feasible.[60] Let us take these heuristics into the world of policy more directly by returning to BRI.

Toward a Twenty-First Century Anthropocene Material Cultural Diplomacy

BRI can trot down the old path of hard/soft power, but China now has an opportunity to draw upon the resources of rhetorical and poetic traditions. In this chapter, we argue that the Belt and Road can advance positions that transcend the techniques of propaganda for national interest, especially under the stresses of climate change. What if a continental power drew upon cultural resources, poetic imagination, in naming ties with island peoples and constructing a symbolic belt that links common efforts to sustain life or prepare for mobility? Part of the idea, here, is to blend the ingredients of *verum* (truth) and *mythos* (falsum) into a *plasma* (a cognitive field of floating symbols) that entertains audiences in plans across remote distances, the curved spaces of the planet. Empathy is bounded only by the limits of moves that call appreciation of common precarious conditions and vulnerabilities into being.

For Chinese diplomacy within the South Pacific, BRI is the text on which new relationships can be worked out. Patricia Riley and colleagues address the role of narratives and scenarios in strategic communication, as humans are "storytelling animals."[61] They show how organizations use strategic communication in the format of campaigns. Likewise, diplomacy requires strategic maneuvering, but the island-state context poses novel demands for justice, given that continents pollute, and islands, in turn, endure the consequences. Thus, the network of spaces created by narrative and material flows must address the precarious urgencies posed by inescapably material conditions, existential events, and threats of irreparable damage. The challenge to CD, then, is to speak to these material facts while crafting a compelling narrative to address what *is* while imagining what *might be*.

The orientation of cultural diplomacy points toward negotiating common action in anticipation of climate change, moving beyond the continental bias. Islands and continents are linked by the irreparable. Rhetorical mapping links with geographical mapping, unveiling the romantic imagination of the remote area, the fantasy of the islands. Poetics and diplomacy, together, impart hope for transforming the island imaginary into the truth of impending change, beyond the leagues of distance. From this perspective, we propose to understand the BRI as a discourse linking continents and islands. In particular, the Maritime Silk Road part of BRI puts into the realm of the possible cooperative opportunities between China and the South Pacific Islands two biomes connected by a common environmental jeopardy—climate change.

We accordingly address each island's specific ecologies and relations with China. Their positions fall into three narratives: Fiji as *historia*, long having had diplomatic relations with China; the Marshall Islands as *plasma*, without deep diplomatic ties to China currently, though one might easily imagine developing them as feasible; and Tuvalu as *mythos*, stretching into the realm of fantasy and fairy tales, insofar as it appears not to have the possibility of benefiting from Chinese cultural diplomacy. Although the turn to such stories might feel uncomfortable in these life or death contexts, Dipesh Chakrabarty argues that the climate crisis is one of imagination.[62] We too believe these narratives might help us develop a more robust cultural imaginary of our possible futures.

To elaborate, in these pages we hope to animate the following tales. To begin, we position Fiji as the representative of *historia*, the friend. It is the first country in the South Pacific to build official diplomatic relations with China. It is also the most welcoming of President Xi's invitation to join BRI among countries in that region. In fact, it is the only South Pacific country to join BRI formally thus far.[63] In other words, the cooperation between China and Fiji has been formed historically as a fact. In contrast, we argue that Tuvalu represents *mythos*, the "hostile," because it does not have diplomatic relations with China. Moreover, among all South Pacific countries not having official diplomatic relations with China, it is the only one whose trade relations are declining—both import and export are dropping year by year.[64] The cooperation is still a fantasy. Third, the Marshall Islands represent *plasma*, the neutral and the possible, because politically, the country turned to Taiwan and terminated diplomatic relations with China, but economically, it is the largest trade partner of China in this region.[65] To China, the Marshall Islands are a *plasma*. They are not as close as Fiji, but also not as disconnected as Tuvalu. As a country in between, they have more possibility to transform the cooperation from fantasy to truth.

Fiji as *Historia*

Located in the heart of the South Pacific Ocean, the People's Republic of Fiji consists of three hundred islands and atolls, but only one-third of them have residents. It also has a relatively large population—more than 870,000 people. The archipelago is vulnerable to many disasters such as cyclones and floods.[66] Before Europeans arrived in 1643, Fiji had been inhabited by Melanesians. It became a crown colony of the British in 1874, gained its independence in 1970, and the sovereign government was established in 1987.[67] Fiji is "one of the smallest contributors to global carbon emission," but it is heavily affected by climate change. Prime Minister Frank Bainimarama spoke at the UN Climate Change Conference COP23: "Unless the world acts decisively to begin addressing the greatest challenge of our age, then the Pacific, as we know it, is doomed."[68] Every year, sea level rises by 6 millimeters, meaning that with current trends in global warming, Fiji will likely be underwater in fifty years.[69] Currently, local islanders suffer from extreme weather, stronger El Niño storms, tropical cyclones, ocean acidification, food and water insufficiency, waterborne diseases, and a rather weak industrial system. The residents are migrating to higher lands year by year.[70] During the UN Sustainable Development Summit in 2015, Prime Minister Bainimarama pointed out the severe situations faced by all small island nations: "We are all in the same boat with this problem that is brought about by climate change. We want to tell the people, the international community, that we all need to come on board this journey. Otherwise it's not gonna do us any good."[71] Migration is also on his agenda. Currently, "the nation is investing tens of the thousands of euros in developing a legal framework to help relocate future climate refugees from other Pacific islands countries."[72] But when the water comes, Fiji will be in trouble.

China and Fiji have perhaps the longest diplomatic relations in the South Pacific, though the two countries only established formal relations on November 5, 1975. Since then, dozens of leaders from the two countries have visited each other. President Xi Jinping visited Fiji in November 2014. The president announced that the two nations would establish a strategic partnership with mutual respect and joint development. Trade relations have been robust. China exports machines, seafood, computers, automobiles, ships, and so on, and imports alumina, timber, and frozen fish. China also organized cultural and artistic troupes to Fiji; moreover, China sends teachers and provides scholarships for Fiji students to visit China. In 2014, China sent medical personnel and equipment, and in 2015 the two countries signed a Memorandum of Understanding on the abolition of visa requirements.[73] Taken as a whole, China's support to Fiji has tried to improve residents' living conditions, has enhanced female employment, and has strengthened China-Fiji connections.[74]

Amicable diplomatic relations between China and Fiji have warmed *historia* into bilateral cooperation on climate change, yet China does not do enough.

Chinese aid should be more dedicated to strengthening island adaptation resilience and survival. China, as a continental civilization, has its biases as an assembly of biomes, securing primary attention to its own sustainability. The aid from China to Fiji is problematic in three ways: the sum of funds is vague; the projects were on a governmental level, lacking close connection with local Fijians; and the aid has been used mainly for construction, but the maintenance and upgrading needs to follow.[75] The problems generate questions and opportunities for growing ties with China. Most importantly, China should consider island climate refugees. China has become an immigrant country; however, due to the large domestic population, China's government has not formed complete laws and regulations for migration. The official attitude toward openness does not settle well, with negative reactions from the public.[76] Since Fiji has a close relationship with China, it can be the first example to test the practicability of ecologically driven migration. Timely exits away from catastrophic events are always problematic; before the water finally comes, there is a tendency to stay in place or return. China still has time to cultivate greater empathy in its people toward migrants, and to construct a fair and decent system of reaction. The relocation of climate-change refugees will be an important event faced by the South Pacific, China, and global communities.

Tuvalu as *Mythos*

Tuvalu was once known as Ellice Islands. Located in the middle of the South Pacific Ocean, Tuvalu covers a mere 26 square kilometers. The area was first inhabited by Polynesians. It is one of the smallest and most isolated places in the world.[77] Because of its scarce resources and barren land, the local economy largely trades on fishing and tourism.[78] Tuvalu does not have any rivers; therefore, access to fresh water is a serious challenge.[79] In 1568, the Spanish were the first Europeans to arrive in the region. In 1892, the British colonized the Ellice Islands as its protectorate; in 1976, Ellice Islands colony was divided into the British colonies of Kiribati and Tuvalu. Then, in 1978, after almost one hundred years of colonization, independence was negotiated for both Kiribati and Tuvalu.[80] Throughout these years, China and Tuvalu never built official diplomatic relations; rather, following the urging of the United States, Tuvalu established relations with Taiwan in 1979. Because of this choice, the volume of trade between China and Tuvalu has been dropping. In 2016, the number was 5.65 million US dollars, a decrease over the prior decades of 65 percent.[81] Communication between the two countries is also insignificant, as seen in the fact that less than fifty Chinese visited Tuvalu in 2017.[82] Regardless of its political affiliations, the environmental situation is extremely precarious in Tuvalu. The highest point on the island is only about 4.5 meters above sea level, but the greatest swell striking this country can be 3.2 meters high.[83] Given the current projections of how climate change will lead to raised sea levels, Tuvalu's future, pending drastic action, will be underwater.[84]

Hence, when the prime minister of Tuvalu, Enele Sopoaga, spoke at COP21 in 2015, he said, "If we save Tuvalu, we save the world." The announcement was not a complete overstatement. At that year, Tuvalu was hit by Cyclone Pam. Although it has received foreign aid, the tremendous damage caused by the single disaster, and the continuous influence of climate change, requires more attention, awareness, and action. The prime minister of Tuvalu advocates that the world should jointly mitigate climate change and protect human rights, which goes beyond national interests and bias.[85] Under such harsh circumstances, Tuvalu people face even more difficulties in terms of their status. In 2017, two Tuvalu families were rejected as refugees because the climate-change crisis does not meet the conventional definitions for asylum seekers.[86] To build refugee-status recognition is an urgent call for climate-change consequences.

For Tuvalu, the *mythos*, breaking the deadlock can be a starting point. Is there any way to get beyond the diplomatic hostility, gathering available resources as much as possible to deal with the common problem together? *Kairos*, the timing, may be essential. When the water finally comes, will humanity be the priority, whatever the politics? We must value survival first before thinking about politics. The imagination of Tuvalu's fate and the globe's response to its people prompt a diplomacy that foresees shores, waters, and shared destinies. Facing climate change, will the country become an opportunity for emerging powers such as China to cooperate more substantially with others? Will climate change, the humanitarian issue, punctuate geopolitics? Will BRI be flexible and inclusive enough to navigate among different interests? The story of Tuvalu is the place of myth, a remote space where the fate of a world is being worked out. It is an urgency calling for China to imagine the way into evolving diplomacy between human communities in the Anthropocene. And here, most importantly, where US influence is strong, is where our shared futures may evolve, as the next generation of green communication and care will be negotiated by the people of Tuvalu, all while situated between China and the United States.

Marshall Islands as *Plasma*

Inhabited by more than fifty thousand people, the Marshall Islands are comprised of twenty-nine coral limestone and sand atolls and five islands. These are scattered in 750,000 square miles of ocean area.[87] The Republic of the Marshall Islands is a US-associated state and has a complicated history and relations with the United States. In the 1940s and 1950s, the US tested fifty-four nuclear bombs in the Bikini atoll. Although the United States paid $604 million to "repair" the situation, the money has been used up. Alongside the physical damage caused by the bombing tests, the cultural legacies may be even more difficult to overcome.[88] China's relations with the Marshall Islands swing back and forth. Diplomatic relations were established on November 16, 1990, but in 1998, China terminated the relation because the Marshall Islands established diplomatic exchanges with Taiwan. Economically, the Marshall Islands are the largest trading partner with China in the South Pacific, yet the relationship illustrates why China's trade practices have been so thoroughly criticized globally, as exports to the island are increasing while imports from the island are decreasing—hence creating a devastating trade imbalance in favor of the continental power over the island.[89]

The water encroaches onto the habitable lands. Meanwhile, the island chain is tortured by extreme weather. Christopher Loeak, the former president of the Republic of the Marshall Islands, described the situation this way: "In the last year alone, my country has suffered through unprecedented droughts in the north, and the biggest ever king tides in the south; and we have watched the most devastating typhoons in history leave a trail of death and destruction across the region."[90] As a consequence, the local fishery industry and tourism have been adversely affected. The Majuro atoll, home to half of the nation's residents, would lose 80 percent of its land if sea levels rise by one meter.[91] To borrow Cox's notion, then, the Marshall Islands are the living embodiment of *the precarious*.

This is why some Marshall Islands frontline communities have reiterated the goal of reducing "1.5 to stay alive," a demarcation for death and survival, in contrast to the 2 degree goal that has prevailed to date at global climate negotiations.[92] The islands will disappear within fifty years if temperatures increase at current rates.[93] What is worse, the nuclear waste left by the United States has become "a poison."[94] Rising seas are leaking into the buried nuclear waste and spreading the poisonous elements, threatening residents' lives. Schoolchildren attend classes near waste sites because of the limited land on Enewetak atoll. Local people are vulnerable to food, water, and soil contamination. The US Department of Energy has even forbidden importing fish and copra from the region. Once serious leaks occur, not only the Marshall Islands, but the whole Pacific will be contaminated. And so the Marshall Islands ascend as a symbol that fuses the nuclear age with climate change. Moreover, the Marshall Islands' irreparable situation has led to an increasing number of climate-change migrants, as more than ten thousand people have been relocated to Hawaii, Washington, and Arkansas.[95] The Marshall Islands thus bring together three of the great challenges of the age: reckoning with Cold War nuclear damage, addressing climate change, and figuring out ethical responses to global migration chains.

In the Marshall Islands' case, the major factor influencing its relationship with China is Taiwan. To simply ignore politics and then to approach environmental problems may be naive, but to let politics obstruct cooperation environmentally is short-sighted. With the current US administration, these islands are not likely to obtain more resources. China's attitude and actions, therefore, become even more important than they might be otherwise. Indeed, particularly given the significance of US-Taiwan ties, for China to imagine a way to work peacefully with the Marshall Islands might indicate a more regional approach, one where the United States, Taiwan, China, and other Oceania states learn to put aside the politics of rivalry in favor of an ethic of care. Thus, for China, the Marshall Islands might best be imagined as *plasma*, a place that is remote and holds possibilities for how to prepare an island to mitigate the impacts of immanent disasters exacerbated by climate change. The process to bolster China's cooperation with the islands requires overcoming or temporarily setting aside ideological contestation in order to act together to prevent the worst imaginable impacts of climate change. A window of trade exchange could be opened, since the trade volume is significant enough. More resources can be provided for residents to help them live a decent life before evacuating. Then the economic-related questions such as investment can be further explored. Referring to former experience, negotiations for more facilities and infrastructure have begun to build climate-change warning and supervising systems to help save vulnerable lives.[96] The Marshall Islands can become part of global warning networks, which provide information for other islands to share.

Conclusion

China may be on its way toward building a twenty-first-century cultural diplomacy of the Anthropocene, yet the Belt and Road Initiative faces material questions. Critique is necessary to rethink the factors that make up diplomatic engagement among remote, asymmetric, vulnerable, and interest-driven states and peoples. And so, drawing on ancient rhetorical heuristics, we identified three Oceania nations entangled in distinctive land-island narratives. First, we approached Fiji as an exemplar of BRI material cultural diplomacy, and thus as a case study of what we have called *historia*. In this case, the facticity of climate refugees will require further negotiation of land-island relations with and beyond Fiji. This human rights crisis appears to be looming in our near future as a result of climate change. We then considered Tuvalu as a *mythos* stretching its present story into a fantasized future. The plausible time of "sinking" unfolds a tale of the impossible: imagining a nation covered by water, disappearing without a trace. This crisis raises concerns of climate-refugee human rights as well as how we might remember a nation that is on the verge of disappearing. Finally, we depicted the Marshall Islands as *plasma*, a fictively feasible tale. We argue that the BRI calls to imagine the fate of the world intertwined with the fate of this island chain. Comparison provides a generative way to bring into view the global work that needs to be done: from reducing greenhouse-gas emissions, to addressing climate-refugee rights, to mitigating immanent disasters, such as by installing improved warning, evacuation, and recovery systems on the islands. Across these three case studies, we emphasized the power of imagination, of using present crises as prods to try to envision new ways of living together across national histories and political rivalries.

Before concluding, we should note that Oceania is not alone. The three islands discussed here represent the likely fate for many other water-surrounded regions. Untold numbers of islands exist across the planet, and eleven thousand are inhabited.[97] The Caribbean, Indian Ocean, and Mediterranean peoples, for instance, all have shared goals: to survive, working through the changing, multiple consequences of climate change on season, locality, and environment.[98] For example, Torres Strait Islanders recently sued Australia for failing to reduce carbon emissions. Their argument is that the loss of their islands is the loss of their way of life and culture and, thus, fundamentally violates their human rights.[99] This example demonstrates the urgency that islanders are encountering globally as they face precarious spaces and irreparable transformations. Water separates these biomes, true; but it also connects them with waves of hazards, possibility, and imagination. This water orientation challenges the land-centric bias in communication research by "remapping . . . place in creative and dynamic connections with land and ocean."[100] Over time, the mix of development, tourism, migration, and resilience will come to create a legacy of events that continue to generate new challenges to island ways of life. As we have shown here, material diplomacies of the Anthropocene shall be practiced, whether successful or not, and they will take on their own generative ambits, feeding not only our evolving notions of green communication, but the fabric of US-China relations in Oceania and beyond.

NOTES

1. Usman W. Chohan, "What Is One Belt One Road? A Surplus Recycling Mechanism Approach," *SSRN E-library* (July 2017): 1, http://dx.doi.org/10.2139/ssrn.2997650.

2. Ibid.; J. P., "What Is China's Belt and Road Initiative," *The Economist*, May 15, 2017, 1.

3. Belt and Road Portal, "The Belt and Road Initiative Progress, Contributions, and Prospects," April 22, 2019, https://eng.yidaiyilu.gov.cn/zchj/qwfb/86739.htm.

4. Joachim H. Spangenberg, "China in the Anthropocene: Culprit, Victim or Last Best Hope for a Global Ecological Civilisation," *BioRisk* 9 (June 2014): 1.

5. Chao Zhang, "Why China Should Take the Lead on Climate Change," *The Diplomat*, December 14, 2017, 1, https://thediplomat.com.

6. Terence Wesley-Smith, "China's Rise in Oceania: Issues and Perspectives," *Pacific Affairs* 86, no. 2 (June 2013): 351–372.

7. AOSIS and UNDP, "Rising Tides, Rising Capacity: Supporting a Sustainable Future for Small Island Developing States," June 2017, https://www.undp.org.

8. Chris Ballard and Bronwen Douglas, *Foreign Bodies: Oceania and the Science of Race, 1750–1940* (Canberra, Australia: ANU Press, 2008).

9. "Countries in Oceania," Worldometers, https://www.worldometers.info/geography/how-many-countries-in-oceania.

10. Brian J. Hurn, "The Role of Cultural Diplomacy in Nation Branding," *Industrial and Commercial Training* 48, no. 2 (2016): 80.

11. Milton C. Cummings, *Cultural Diplomacy and the United States Government: A Survey* (Washington, DC: Center for Arts and Culture, 2003), 1; Hurn, "The Role of Cultural Diplomacy in Nation Branding," 80.

12. Joseph S. Nye, *Soft Power: The Means to Success in World Politics* (New York: Public Affairs, 2004), 5. See also Patricia M. Goff, "Cultural Diplomacy," in *Oxford Handbook of Modern*

Diplomacy, ed. Andrew F. Cooper, Jorge Heine, and Ramesh Thakur (Oxford: Oxford University Press, 2013), 1; Hyungseok Kang, "Contemporary Cultural Diplomacy in South Korea: Explicit and Implicit Approaches," *International Journal of Cultural Policy* 21, no. 4 (June 2015): 433–447.

13. Tae Young Kim and Dal Yong Jin, "Cultural Policy in the Korean Wave: An Analysis of Cultural Diplomacy Embedded in Presidential Speeches," *International Journal of Communication* 10 (2016): 5529; Hurn, "The Role of Cultural Diplomacy in Nation Branding," 80; Melissa Nisbett, "New Perspectives on Instrumentalism: An Empirical Study of Cultural Diplomacy," *International Journal of Cultural Policy* 19, no. 5 (July 2012): 557.

14. Goff, "Cultural Diplomacy," 5.

15. Neil Collins and Kristina Bekenova, "European Cultural Diplomacy: Diaspora Relations with Kazakhstan," *International Journal of Cultural Policy* 23, no. 6 (July 2017): 733.

16. Nisbett, "New Perspectives on Instrumentalism," 571.

17. Goff, "Cultural Diplomacy," 1; David Clarke, "Theorising the Role of Cultural Products in Cultural Diplomacy from a Cultural Studies Perspective," *International Journal of Cultural Policy* 22, no. 2 (September 2014): 160.

18. Hermann F. Eilts, "Diplomacy—Contemporary Practice," in *Modern Diplomacy: The Art and the Artisans*, ed. Elmer Plischke (Washington, DC: American Enterprise Institute for Public Policy Research, 1979), 5.

19. Hugh S. Gibson, "Secret vs. Open Diplomacy," in Plischke, ed., *Modern Diplomacy*, 129–134.

20. Yudhishthir Raj Isar, "Cultural Diplomacy: India Does It Differently," *International Journal of Cultural Policy* 23, no. 6 (June 2017): 706.

21. Hurn, "The Role of Cultural Diplomacy in Nation Branding," 80–85; Goff, "Cultural Diplomacy," 5; Isar, "Cultural Diplomacy: India Does it Differently," 711–716; Gay McDonald, "Aboriginal Art and Cultural Diplomacy: Australia, the United States, and the Culture Warriors Exhibition," *Journal of Australian Studies* 38, no. 1 (December 2013): 1–14; Manuel R. Enverga III and Maria Ysabel A. Tangco, "Cine Europa: Behind the Scenes of a Collaborative Cultural Diplomacy Initiative in the Philippines," *International Journal of Cultural Policy* 23, no. 1 (December 2017): 1–14; Kim and Jin, "Cultural Policy in the Korean Wave," 5517–5529; Kang, "Contemporary Cultural Diplomacy in South Korea," 433–447; Vincent Kuitenbrouwer, "Beyond the 'Trauma of Decolonisation': Dutch Cultural Diplomacy during the West New Guinea Question (1950–62)," *Journal of Imperial and Commonwealth History* 44, no. 2 (2016): 306–327; Miia Huttunen, "De-demonising Japan? Transitioning from War to Peace through Japan's Cinematic Post-war

Cultural Diplomacy in UNESCO's Orient Project 1957–1959," *International Journal of Cultural Policy* 23, no. 6 (2017): 752; Martina Topić and Cassandra Sciortino, eds., *Cultural Diplomacy and Cultural Imperialism: A Framework for the Analysis* (Bern: Peter Lang, 2012), 14; Nisbett, "New Perspectives on Instrumentalism," 557; Xin Xin, "Chindia's Challenge to Global Communication: A Perspective from China," *Global Media and Communication* 6, no. 3 (2010): 296.

22. Kang, "Contemporary Cultural Diplomacy in South Korea," 433.

23. Herbert Passin, *China's Cultural Diplomacy* (New York: Praeger, 1963), 1–10, 17–25.

24. Ibid., 1, 9–10.

25. Ankit Panda, "Reflecting on China's Five Principles, 60 Years Later," *The Diplomat*, June 26, 2014, https://thediplomat.com.

26. Wanning Sun, "Slow Boat from China: Public Discourses behind the 'Going Global' Media Policy," *International Journal of Cultural Policy* 21, no. 4 (June 2015): 400–418; Wanning Sun, "Configuring the Foreign Correspondent: New Questions about China's Public Diplomacy," *Place Branding and Public Diplomacy* 11, no. 2 (June 2014): 127; Ien Ang, Yudhishthir Raj Isar, and Phillip Mar, "Cultural Diplomacy: Beyond the National Interest," *International Journal of Cultural Policy* 21, no. 4 (June 2015): 372, 373.

27. Sun, "Configuring the Foreign Correspondent," 127; Sun, "Slow Boat from China," 400–418.

28. Hongying Wang, "China's Image Projection and Its Impact," in *Soft Power in China*, ed. Wang Jian (New York: Palgrave Macmillan, Series in Global Public Diplomacy, 2011), 43; Sun, "Slow Boat from China," 400–418.

29. Reality Check Team, "Reality Check: Is China Burdening Africa with Debt?," *BBC*, November 5, 2018, https://www.bbc.com; John Hurley, Scott Morris, and Gailyn Portelance, "Examining the Debt Implications of the Belt and Road Initiative from a Policy Perspective," Center for Global Development, March 4, 2018, www.cgdev.org.

30. Bill Willmott, "Varieties of Chinese Experience in the Pacific," CSCSD Occasional Paper no. 1 (2017): 36–37; Jing Wang, "Zhongguo shouge yanjiu taipingyang daoguo zhiku fabu, jiang zhongdian guanzhu yidaiyilu jianshe [China launched the first think tank studying Pacific Islands, which will focus on OBOR initiative]," *China.com* (Beijing), September 7, 2017, http://news.china.com/news100/11038989/20170907/31312202_all.html.

31. "Haishang sichou zhilu qiannian xingshuai shi [The wax and wane of Maritime Silk Road history]," *People's Daily* (Beijing), May 20, 2014, http://history.people.com.cn.

32. Renxia Chang, *Haishang sichou zhilu yu wenhua jiaoliu* [Maritime Silk Road and cultural communication] (Beijing: Beijing Publishing Group, 2016), 1–46.

33. Xiujun Xu, "China and the Pacific Island Countries," *Contemporary International Relations* 20, no. 4 (July/August 2010): 127.

34. Joel Atkinson, "China–Taiwan Diplomatic Competition and the Pacific Islands," *Pacific Review* 23, no. 4 (August 2010): 407–427.

35. "28 guo lingdaoren jiang chuxi yidaiyilu luntan xijinping zhuchi fenghui [28 countries are going to participate in One Belt One Road Forum and President Xi is going to host it]," Caixin Web, April 18, 2017, China, http://china.caixin.com; "China Woos Pacific Nations for Silk Road," *SBS News*, May 29, 2017, https://www.sbs.com.au.

36. "Nan taipingyang yu haishang sichou zhilu [South Pacific and Maritime Silk Road]," Ministry of Commerce of the People's Republic of China, May 23, 2017, 1, http://cafiec. mofcom.gov.cn/article/tongjipeixun/201705/20170502581806.shtml; Xiaohong Zheng, "Jingji Xuejia Fangang: zhongguo yu taipingyangdaoguo hezuo zhengdangshi [Economist Fangang: It is the right time for China and South Pacific Islands to cooperate]," *China News* (Beijing), July 13, 2017, http://www.chinanews.com.

37. Qizhi Zhang and Dayong Li, "Feiji zongli: Yidaiyilu wei nantai diqu fazhan dailai xin jiyu [OBOR brings new opportunities for the development of South Pacific Region]," *China Radio International*, May 11, 2017, http://news.china.com/news100/11038989/20170511/30 514656_all.html.

38. "Nan taipingyang yu haishang sichou zhilu [South Pacific and Maritime Silk Road]," 1.

39. Jieqiu Liu and Xingwei Huang, "Zhongguo jiang anpai 2 yi yuan bangzhu taipingyang daoguo yingdui qihou bianhua [China is going to provide 200 million RMB to help Pacific Islands face climate change]," *Qingdao Daily* (Qingdao), September 2, 2012, http:// world.huanqiu.com.

40. Jonathan Gregory, "Projections of Sea Level Rise," in *Climate Change 2013: The Physical Science Basis*, ch. 13 (Geneva: Intergovernmental Panel on Climate Change, 2013).

41. Brian Merchant, "First Official Climate Change Refugees Evacuate Their Island Homes for Good," *Treehugger*, May 8, 2009, http://www.treehugger.com.

42. Angela Dewan, "Five Solomon Islands Swallowed by the Sea," *CNN*, May 10, 2016, http:// www.cnn.com.

43. UNDP, *China's South-South Cooperation with Pacific Island Countries in the Context of the 2030 Agenda for Sustainable Development, Series Report: Climate Change Adaptation*, April 2017, http://www.cn.undp.org.

44. Associated Press, "Climate Change Threatens to Wipe Some Islands off the Map," *Washington Post*, June 23, 2017.

45. Mark Abadi, "These Island Nations Could Be Underwater in as Little as 50 Years," *Business Insider*, December 30, 2015, 1, https://www.businessinsider.com.

46. Notably, Australia is also a significant polluter as well as a neighbor of South Pacific Islands, and it is accused of making insufficient efforts. Australia's emissions are still

increasing, causing island nations to be concerned. Anote Tong, "While My Island Nation Sinks, Australia Is Doing Nothing to Solve Climate Change," *The Guardian*, October 10, 2018.

47. Fiona Harvey, "China Aims to Drastically Cut Greenhouse Gas Emissions through Trading Scheme," *The Guardian*, December 19, 2017.

48. Julie Sze, *Fantasy Islands: Chinese Dreams and Ecological Fears in an Age of Climate Crisis* (Oakland: University of California Press, 2015).

49. Spangenberg, "China in the Anthropocene," 1; "The World Is Losing the War against Climate Change," *The Economist*, August 2, 2018.

50. Sze, *Fantasy Islands*.

51. Ibid.

52. Rebecca Valli, "China Revises Environmental Law," *VOA News*, April 25, 2014, 1, https://www.voanews.com.

53. Sze, *Fantasy Islands*, 35.

54. "'APEC Blue' Tops Beijing Environment Key Words for 2014," *China Daily*, December 24, 2014, 1, http://www.chinadaily.com.cn.

55. Sze, *Fantasy Islands*.

56. Peter Sloterdijk, "Talking to Myself about the Poetics of Space," *Harvard Design Magazine*, no. 30 (2009): 1.

57. Luis Castro Nogueira, "Bubbles, Globes, Wrappings, and Plektopoi: Minimal Notes to Rethink Metaphysics from the Standpoint of the Social Sciences," *Environment and Planning D: Society and Space* 27, no. 1 (2009): 88.

58. Jeff Goodell, *The Water Will Come* (Boston: Little, Brown and Co., 2017), 166, 168.

59. J. Robert Cox, "The Die Is Cast: Topical and Ontological Dimensions of the Locus of the Irreparable," *Quarterly Journal of Speech* 68, no. 3 (August 1982): 229–234.

60. Charles Oscar Brink, *Horace on Poetry* (Cambridge: Cambridge University Press, 1963), 354, 355.

61. Patricia Riley, Rong Wang, Yuehan Wang, and Lingyan Feng, "Global Warming: Chinese Narratives of the Future," *Global Media and China* 1, no. 1–2 (2016): 12–31; Walter R. Fisher, "Narration as a Human Communication Paradigm: The Case of Public Moral Argument," *Communication Monographs* 51, no. 1 (March 1984): 5.

62. Dipesh Chakrabarty, "The Climate of History: Four Theses," *Critical Inquiry* 35, no. 2 (2009): 197–222.

63. "28 guo lingdaoren . . . [28 countries are going to participate . . .]"; "China Woos Pacific Nations for Silk Road."

64. "Zhongguo tong Feiji de guanxi [China's relation with Fiji]," Ministry of Foreign Affairs

of the People's Republic of China, updated August 2017, http://www.fmprc.gov.cn/chn// gxh/cgb/zcgmzysx/yz_1/1206_3/1206x1/t7462.htm.

65. "Zhongguo tong Mashao'er qundao de guanxi [China's relation with Marshall Islands]," Ministry of Foreign Affairs of the People's Republic of China, updated August 2017, http://www.fmprc.gov.cn/web/gjhdq_676201/gj_676203/dyz_681240/1206_681492/ sbgx_681496.

66. United Nations Climate Change, "How Fiji Is Impacted by Climate Change," February 9, 2017, https://unfccc.int/news/how-fiji-is-impacted-by-climate-change.

67. "Fiji: History," The Commonwealth, http://thecommonwealth.org/our-member-countries/fiji/history.

68. COP23, "How Fiji Is Affected by Climate Change," UN Climate Change Conference, BONN, 2017–2018, 1, https://cop23.com.fj/fiji-and-the-pacific/how-fiji-is-affected-by-climate-change.

69. Ibid.; Abadi, "These Island Nations Could Be Underwater."

70. COP23, "How Fiji Is Affected by Climate Change."

71. Josaia Voreqe Bainimarama, "Statement on the Occasion of the UN Sustainable Development Summit," September 26, 2015, https://sustainabledevelopment.un.org/ content/documents/20718fiji.pdf.

72. Sarah Taylor, "Fiji Prepares for 'Climate Refugees,'" 1, *Euronews*, https://www.euronews.com/2017/11/17/fiji-prepares-for-climate-refugees.

73. "Zhongguo tong Feiji de guanxi [China's relation with Fiji]," *Ministry of Foreign Affairs of the People's Republic of China*, updated August 2017, https://www.fmprc.gov.cn/web/ gjhdq_676201/gj_676203/dyz_681240/1206_681342/sbgx_681346.

74. Guixia Lyu, "China's Development Aid to Fiji: Motive and Method," 2015, https://www.victoria.ac.nz/chinaresearchcentre/programmes-and-projects/china-symposiums/china-and-the-pacific-the-view-from-oceania/24-Lyu-Guixia-Chinas-Development-Aid-to-Fiji-Motive-and-Method.pdf.

75. Ibid.

76. Frank N. Pieke, "Emerging Markets and Migration Policy: China," 2014, https://www.ifri.org/en/publications-cerfa-ifri/emerging-markets-and-migration-policy-china-1.

77. Andrea Milan, Robert Oakes, and Jillian Campbell, "Tuvalu: Climate Change and Migration: Relationships between Household Vulnerability, Human Mobility and Climate Change," Report No. 18 (Bonn: United Nations University Institute for Environment and Human Security [UNU-EHS], November 2016), https://collections.unu.edu/eserv/ UNU:5856/Online_No_18_Tuvalu_Report_161207_.pdf.

78. "Zhongguo tong Tuwalu de guanxi [China's relation with Tuvalu]," Ministry of Foreign

Affairs of the People's Republic of China, updated August 2017, http://www.fmprc.gov.cn.

79. Milan, Oakes, and Campbell, "Tuvalu: Climate Change and Migration."

80. "Tuvalu," Wikiwand, https://www.wikiwand.com/en/Tuvalu.

81. "Zhongguo tong Tuwalu de guanxi [China's relation with Tuvalu]."

82. "Tuwalu, yige jijiang chenmo haidi de meili guodu [Tuvalu, a beautiful country which is going to sink under the sea]," *EZDIVE*, October 13, 2017, http://www.sohu.com/a/197975532_232404.

83. Yun Cheng Deng and Yanan Fu, "Tuvalu: qihou bianhua weiji guojia cunwang [Tuvalu: Climate change threatens the survival of the country]," *Zhongguo Haiyang Bao* (Beijing), March 21, 2017, http://www.hycfw.com/Article/203583.

84. Angie Knox, "Sinking Feeling in Tuvalu," *BBC News*, August 28, 2002, http://news.bbc.co.uk.

85. Enele S. Sopoaga, Keynote Statement, COP21, November 30, 2015, 1, https://unfccc.int/files/meetings/paris_nov_2015/application/pdf/cop21cmp11_leaders_event_tuvalu.pdf.

86. G. Bonnett, "Climate Change Refugee Cases Rejected," *RNZ*, October 24, 2017, https://www.radionz.co.nz/news/national/342280/climate-change-refugee-cases-rejected.

87. "Lives in the Balance: Climate Change and the Marshall Islands," *The Guardian*, September 15, 2016.

88. "Marshall Islands," Wikiwand, https://www.wikiwand.com/en/Effects_of_climate_change_on_island_nations#/Marshall_Islands.

89. "Zhongguo tong Mashao'er qundao de guanxi [China's relation with Marshall Islands]."

90. Christopher Jorebon Loeak, "A Clarion Call from the Climate Change Frontline," *Huffington Post*, September 18, 2014, 1.

91. "Marshall Islands," Wikiwand.

92. Ari Shapiro, "For the Marshall Islands, the Climate Goal Is '1.5 to Stay Alive,'" *NPR*, December 9, 2015, 1, https://www.npr.org.

93. Abadi, "These Island Nations Could Be Underwater."

94. Mark Willacy, "A Poison in Our Island," *ABC News*, November 27, 2017, 1, https://mobile.abc.net.au.

95. David Saddington, "Small Islands, Big Impact: Marshall Islands Set Bold Carbon Target," *Huffington Post*, updated December 6, 2017, https://www.huffpost.com.

96. Katie Brown, "New Early Warning System Could Protect Vulnerable Islands from Flooding," *Phys.org*, https://phys.org.

97. World Atlas, "How Many Islands Are There in the World?," https://www.worldatlas.com/articles/how-many-islands-are-there-in-the-world.html.

98. Alliance of Small Island States, "About AOSIS," n.d., http://aosis.org/about-aosis.

99. Livia Albeck-Ripka, "Their Islands Are Being Eroded. So Are Their Human Rights, They Say," *New York Times*, May 12, 2019.

100. Tiara R. Na'puti, "Archipelagic Rhetoric: Remapping the Marianas and Challenging Militarization from 'A Stirring Place,'" *Communication and Critical/Cultural Studies* 16, no. 1 (2019): 2.

Urban Planning as Protest and Public Engagement

Reimagining Mong Kok as an Eco-City

Andrew Gilmore

When hundreds of thousands of people displayed yellow umbrellas while occupying the streets of Hong Kong in late 2014, the city transformed into an environmentally conscious space. For seventy-nine days, Hong Kongers took education, art, farming, religion, and other aspects of everyday life into public spaces to promote a new collective vision. Highlighting the desire to reimagine the future of their city by occupying public space, the protests known as the "Umbrella Revolution" were described by commentators as "the greenest protest in the world."[1] Images viewed across the globe depicted a city where protesters cleared and recycled their litter, designated study corners complete with solar power, composted banana peels to make cleaning fluid, and recycled plastic bottles to make protective face masks.[2]

In a city bursting with over seven million citizens, the Umbrella Revolution showed how Hong Kong's youth—while demanding more democracy, lower housing costs, more transparent media, and a stronger sense of local identity—based their civic participation on a green consciousness. While many Hong Kongers desire democracy, the issue is anything but monolithic in the city.[3] Moreover, many mainland citizens oppose Hong Kong's frequent protests, believing that the city is afforded a host of individual rights that those in the mainland are not. For the Party,

Hong Kong–style democracy in the mainland would lead to "chaos."[4] Considering how this protest was applauded in the US press, it also seems evident that part of US-China relations has hinged on how the US views the roles of environmentalism within China's growing protest culture.[5]

Inspired by the protests, architect Vicky Chan created an urban plan for how Mong Kok—one of the major sites of mass gatherings during the Umbrella Revolution—could lead the way in providing a vision for a more democratic and sustainable Hong Kong. His plan, called "Green Mong Kok" (hereafter GMK), offers a compelling cultural text for considering environmental expressions of the future, as it reimagines democratic possibilities through green urban infrastructure. As the founder of the award-winning Hong Kong–based architectural practice "Avoid Obvious Architects," Chan draws from and pushes the boundaries of the Umbrella Revolution, claiming that GMK "shows elements we learnt from the movement and developed further."[6] GMK is a speculative piece of architecture driven by respect for the protesters and a love of Hong Kong—and hence, I argue, it is a significant artifact for thinking about how to reimagine the role of public spaces in our present and future.

Although the Umbrella Revolution was not a social movement driven primarily by environmental issues, I explore in this chapter the rhetorical appeals of GMK and how it may enable and/or prevent Hong Kongers, to borrow a phrase from Phaedra C. Pezzullo, to "reframe the narratives that sustain oppressive environmental conditions."[7] Chan's work clearly strives to achieve this empowering goal, as he has stated that he hopes his architecture harnesses the Umbrella Revolution to "influence future leaders with creative thinking" about how urban spaces might become more environmentally friendly and more democratic.[8] In this way, Chan situates his work in general, and GMK in particular, within an emerging global discourse in which architects, urban planners, activists, scholars, and city leaders explore ways to build a more sustainable urban future that addresses ecology, economics, and social equity.[9] My analysis suggests, however, that while Chan's GMK offers a stunning future vision for one part of Hong Kong, it also may be infused with lingering notions of class and social hierarchy, which were two of the key injustices that led to the Umbrella Revolution in the first place.[10] Nonetheless, I argue that Chan's urban plan can be interpreted as a tool of protest and as a site of public engagement worthy of attention for environmental communication in China and beyond. Given the fact that China is undergoing what Donovan Conley has diagnosed as the single largest urban migration in the history of the planet, my analysis of GMK serves as a case study in how the future of green communication is being imagined, debated, and lived in Hong Kong, one of the world's megacities.[11]

Studying Space

The everyday physical and performative aspects of a city, posit Greg Dickinson and Giorgia Aiello, lie at the heart of urban communication. The basic materials of a city—bricks and mortar—are communicative because they "shape, constrain, and ultimately also mediate the everyday lives of individuals and communities."[12] In short, as the actions, identities, and practices of the inhabitants of a space are enabled and/or constrained by the communicative modes of the built environment, urban spaces are, suggests Doreen Massey, "the product of power-filled social relations."[13] Taking seriously the "matter, movement, being, and bodies of cities" enables urban communication scholars to produce work that matters and makes a difference.[14] I hope to extend this line of scholarship by arguing that we need to approach the work of urban communication scholars through the lens of environmental communication. Indeed, as "life on Earth is as precarious" as ever before, for Robert Cox, the imperative for studying environmental communication is an ethical one.[15] Bringing together urban communication studies and environmental communication studies is an attempt to bolster both in light of their shared commitments to studying symbolic action, power, and places.

Given the material crises we face in the United States, China, and beyond, it might appear a surprising choice to focus on an artifact that is an imagined space. Agreeing with Dipesh Chakrabarty, however, if the climate crisis is, in part, a crisis of imagination, we must take imaginary projections of potential futures seriously. Our current struggles are not just about where life stands on this planet today, but also what we hope it will become in the future.[16] Studying an imagined space, however, does pose challenges. Not being able to visit GMK means I can only analyze an architectural drawing. My analysis primarily is focused on a close reading of the plan; however, it also reflects my personal experiences in contemporary Mong Kok. During my time living in Hong Kong, I frequented the Mong Kok district on a regular basis.[17] I have walked the streets and smelled its air, including the aroma of stinky tofu from the stall on the corner of Argyle Street and Tung Choi Street, and the smell of orchids from Flower Market Road. I can hear the clacking of the crosswalks as I wait to cross the teeming intersection of Nathan Road and Mong Kok

Road. On afternoons of intense heat and humidity, I have battled through crowds in the Ladies Market to barter for gifts. These experiences, I argue, help me better appreciate the radical transformations Chan is imagining in GMK.

My analysis of this imagined plan also is informed by critical theories of space. Setha Low describes the social construction and production of space as two complementary perspectives for understanding how urban, public space becomes "semiotically encoded and interpreted reality."[18] A social construction framework assesses not just physical properties, but also abstractions that comprise shared understanding and social structural differences. For Low, social and symbolic meanings produced within and by a space draw attention to both transformations and contestations that "occur through social interactions, memories, feelings, imaginations, and daily use, as well as the actions that convey particular meaning."[19] Studying and negotiating democratic public spaces "necessitates operating with a concept of spatiality" that, for Massey, "keeps always under scrutiny the play of the social relations which construct them."[20]

The social production of space should examine at least four facets: social history and development, the political economy, social (re)production and resistance, and social control and spatial governmentality. The social history and development approach "provides a basic understanding of the evolution of architectural and spatial form and reveals its ideological, political, and economic underpinnings." For example, studying the built environment from the perspective of social history and development reveals how society produces buildings that maintain and/or reinforce social forms. Studying the political economy of a space means exploring the "underlying political and economic relations that drive spatial production."[21] This approach aids in explaining the inevitable struggles that exist over the built environment through the reproduction of existing class relations. For Low, the control of the means of production by authoritative powers plays a dominant role in built environments. Defined by Low as "how everyday activities, beliefs and practices, as well as social and spatial structures transmit social equality to the next generation," social production refers to the conditions necessary to reproduce social class.[22] This approach also considers possible resistance to the production and reproduction of space via social and political movements. The study of social control and spatial governmentality within a space, meanwhile, explores the structuring and manipulation of space that communicates social control, as well as the role of power dynamics in the rules of inclusion and exclusion within a space.[23] Despite the intricacies of the four approaches, each methodology "illuminate[s] how a space or place comes into existence" and opens up questions about the neutrality and naturalness of the planning and development of built environments.[24] For Low, each methodological stance—all of which are evident in Chan's design for GMK—furthers the analysis of how historical, political, and economic forces shape the material environment.

In many plans and material spaces, the arrangement and allocation of spaces and objects are "assumed to be transparent . . . but rarely are." To this end, when spaces are critically analyzed through the lens of social construction, "spatial formations and relationships yield insights into unacknowledged biases, prejudices, and inequalities in a particularly forceful way."[25] Michael Peter Smith posits that architects and planners of spaces and structures possess "socially constructed understandings of how the world works," although these understandings and assumptions, which take the forms of "exclusionary practices concerning race, class, and gender," can be implicitly or explicitly embedded in ideas and designs.[26] For the purposes of this chapter, I am intrigued by the rhetorical modes GMK uses to invite its inhabitants to interact with each other, and hence envision a green future for Hong Kong.

Hong Kong as a Quintessential Chinese Eco-City

With an average of 6,690 people per square kilometer, Hong Kong is one of the most densely populated cities in the world.[27] As a result, issues of energy efficiency, pollution control, and overcrowding are all conditions that affect the daily lived experience of Hong Kongers.[28] In fact, upon Hong Kong's return to Chinese rule in 1997, the Government of the Hong Kong Special Administrative Region (GovHK) attempted to address the city's environmental challenges. Developed "in response to the need to take account of [the] environmental and social concerns" of Hong Kong, the GovHK commissioned a Study on Sustainable Development for the 21st Century in Hong Kong (SUSDEV 21), which aims to achieve the goal of creating a sustainable, urban-based way of living for the city's residents.[29] Despite these intentions, however, a report published at the turn of the millennium by the University of Hong Kong's Center of Urban Planning and Environment Management revealed that the city was increasingly moving away from sustainable development.[30]

With the ninth largest "ecological deficit" in the world, in 2013, the World Wildlife Fund (WWF) and the Global Footprint Network reported that if everybody in the world lived like Hong Kongers, almost three Planet Earths would be needed to satisfy global demand for resources. Hong Kong's "ecological deficit"—determined by the gap between "ecological footprint" (carbon footprint, plus the demand for resources such as crops and land) and "biocapacity" (the supply of biologically productive land and sea area)—was so bad that the city's demand for resources was 150 times greater than supply.[31] An overreliance on coal power, a lack of renewable energy resources, poor water and air quality, terrible amounts of light pollution, and spiraling amounts of plastic waste mean that Hong Kong is "lagging far behind international standards" with regard to environmental initiatives and emission reduction.[32] The World Health Organization (WHO) also provides damning statistics. Levels of cancer-causing pollutants in Hong Kong have exceeded WHO standards for almost twenty years, and pollution levels are more than four times WHO recommendations. In 2016, air pollution reportedly led to the premature death of over 1,600 people in Hong Kong, and in January 2017, over 300,000 doctor's visits were attributed to smog.[33] As these reports indicate, there is good reason for environmentalists to look to Hong Kong.

Social inequity also poses challenges to Hong Kong. In a megalopolis that houses the sixth highest number of billionaires in the world, materialistic wealth is beyond the realm of possibility for a majority of Hong Kong's residents.[34] When judged alongside nations with whom GovHK compares the city's socioeconomic performance (including mainland China, the UK, and the United States), Hong Kong's income distribution is the most uneven.[35] The huge income disparity between the wealthiest and poorest 10 percent of the city's population is highlighted by the fact that the richest households earn around 44 times that of the poorest households.[36] The ever-increasing cost of living, coupled with the visibility of a growing wealth disparity across the city lays bare the deep socioeconomic and class schism in the city and, in turn, has led to concerns that Hong Kong has become a mainlandized plutocracy controlled by elite, wealthy members of Hong Kong and Chinese society.[37] Hong Kong, of course, is not the only city in which well-being and want coexist. But, with the combined income in over 50 percent of the city's households standing at a little over US$3,000 per month, and the average Hong Kong home costing around 19 times the median salary, the situation in Hong Kong is particularly startling.[38] Despite government attempts to address Hong Kong's vast wealth inequality, the shortage of affordable dwellings is a major issue in what is the most expensive—and one of the least equitable—housing markets in the world.[39]

For many scholars, eco-civilizations—primarily characterized by a focus on a "harmony between humanity and nature"—are a necessary path to stemming the global ecological crisis. As we evolve toward eco-civilizations, the concept of eco-cities in which the city and its inhabitants form "an organic whole" has become especially prevalent.[40] Home to the biggest population on Earth, China is more pressed than most nations to find a solution to its ecological issues. But as a nation with superpower ambitions, how is China faring in addressing some of its major environmental challenges?

As China's economy and population continue to grow, the relentless increase of its energy consumption is unsustainable. Issues including desertification—which has already claimed 27.3 percent of China's total land area, and continues to grow by almost 2,500 km^2 per year—mean that China's ecosystems have become "exceptionally vulnerable." This desertification has led to an influx of internal migration as residents are forced to leave their towns annually. By 2030, experts suggest that approximately one billion people will live in Chinese cities.[41] In light of these local ecological issues and the global climate crisis, much scholarly attention has been paid to China's increasing development and promotion of green policies, many of which act as forms of social advancement designed to "improve the inherent quality of urban life as a whole."[42] In 2014, Chinese Premier Li Keqiang announced that China planned to "resolutely declare war against pollution."[43] Since then, China has introduced a number of green initiatives, including plans to dismantle coal-fired power plants, improve water quality, impose taxes on polluters, and allocate 123.4 trillion RMB (US$19.4 trillion) to help finance green economies.[44] Four years after Li's proclamation of war, reports suggest that China has begun making progress. For example, concentrations of fine particulates in Beijing's notoriously filthy air have reduced by 35 percent, and Baoding—reported to be China's most polluted city—has seen a 38 percent reduction.[45] Despite its recent drive towards environmental improvements, however, China's battle against pollution is not new.

Since the 1980s, China's widespread plans to build "harmonious cities" have seen the conversion of rural land into "new cities" or "satellite towns," which have slowly morphed into being termed eco-cities or "smart cities."[46] As China has continued to promote sustainable development via its plan to move up to 300 million citizens from rural to urban locations, hundreds of Chinese cities have announced intentions to become "eco-cities."[47] The rush to urbanize was so rapid that between 2011 and 2013 China used more concrete than the United States used in the entire twentieth century.[48] In fact, the overproduction of eco-cities has been problematic. As China looks to "international best practices" as a way of transforming its municipalities into "benchmark cities in the global arena," overly ambitious visions, unrealistic goals, and ineffective policy instruments have exposed the shortcomings in China's green credentials.[49] This rapid—and possibly unneeded—development has led to the appearance of hundreds of so-called "Ghost Cities" across the country. Despite possessing every amenity associated with a city, including high-rise apartments, metro stations, and public parks, the swell of new cities are lacking a vital ingredient: residents.

While concrete definitions of the goals and aspirations of eco-cities are difficult to define, Martin de Jong, Dong Wang, and Chang Yu explain them by way of differing patterns for producing and consuming energy, eliminating motorized traffic, and relying on local resources. In short, eco-city aspirations are in direct contrast to the "prevailing consumer societies" that dominate our current global desires.[50] When considering the inhabitants of these eco-spaces, the major goal of an eco-civilization and eco-cities is to prioritize the interests of the entire human race, while advocating for "global governance and world citizenry" and justice between current and future generations.[51] Ideally, then, eco-cities are meant to embody green communication by literally building new spaces for sustainable justice to flourish.

The Architect behind Green Mong Kok

Vicky Chan is an award-winning architect specializing in interiors, landscape, and urban planning, with a major emphasis on sustainable architecture.[52] To showcase one of his many designs for environmentally friendly architecture, in 2015, Chan took his design for GMK to Cape Town, South Africa. By doing so, Chan aimed to inspire future leaders to "think creatively" via the intersection of "art" and "freedom of speech."[53] This notion of free speech, however, is problematic in China. Chan admits that, over the years, his website has been taken down on various occasions by the Party. Despite this, however, his decision to return to Asia from the United States was influenced by his belief that China is more willing to invest in and take risks on young designers.[54] Perhaps, then, there is hope for public engagement in China.

But how did the Umbrella Revolution lead to Chan imagining GMK? For Chan, it was "a once in a lifetime event that [he] had to document." The environmental impact of the protests was the first of two elements of the movement that Chan sought to harness in his design for GMK. Although Chan did not participate in the on-street occupation, his curiosity was piqued by reports of protesters building wind turbines to charge cell phones. With Hong Kong's streets transformed into traffic-free spaces, Chan was struck by the improvement in breathable air, which, for him, transformed Hong Kong into "a utopia." This sense of a growing green utopia was the catalyst behind the second element of the Umbrella Revolution that Chan sought to replicate in GMK: the social aspect of the movement. "Store owners actually knew their customers by name," Chan says. "They were offering discounts." Moreover, "people were sleeping safely on the streets, everybody was living together." By living communally on the streets of Hong Kong for almost three months, spontaneous interpersonal interactions occurred with greater frequency. Conversations among different sections of Hong Kong society led to the visibility of diversity in Hong Kong. For Chan, the promotion of diversity is "the only way for a society to grow, but it is often lacking in Hong Kong." Both green protest practices and socially just public engagement, therefore, were motivations behind the intent of Chan's GMK.[55]

Through GMK, Chan professes to maintain "the core and heritage of Mong Kok."[56] To this end, GMK aims to comprise elements of the district's historical past, aspects harnessed from the present, and visions of a green and just future. Among all of these categories, Mong Kok is not only the busiest but also one of the most polluted districts in Hong Kong.[57]

Mong Kok Today

With over 130,000 people jammed into each square kilometer, Mong Kok—meaning "busy corner" in Cantonese—is one of the most densely populated places on the planet.[58] Sitting on the Kowloon side of Hong Kong and less than an hour's public transport ride from the mainland Chinese city of Shenzhen, Mong Kok is a major transportation hub served by multiple subway lines, overground train lines, and countless bus and minibus routes, all of which contribute to Mong Kok's filthy air.[59] Despite GovHK attempts to reduce air pollution by decreasing traffic, the exhaust produced by Mong Kok's three thousand restaurants and food shops is, in some cases, more harmful to residents than vehicle pollution.[60] Additionally, the district's abundance of high-rise buildings complicates the dispersal of dangerously high and harmful levels of air pollution.[61]

Just a few short blocks from the intersection of Sai Yeung Choi Street and Nelson Street—one of the major sites of mass gathering during the Umbrella Revolution—sits one of Mong Kok's many markets. Here, market stalls where shoppers are expected to haggle for a bargain tightly line both sides of the pedestrianized street. The stalls sell everything from clothing and cellphones to Chinese tea sets and "Armani" luggage. Directly behind the market stalls, permanent stores sell similar goods at similar, but less negotiable, prices. Between the permanent stores, iron gates reveal doors that hide endless staircases leading to tiny apartments, brothels, pool halls, hotpot joints, and massage parlors of all varieties. On a bright summer day, walking through the market can be disorienting as "shoulder to shoulder" buildings block sunlight. Above the market stalls, neon signs reach out from buildings on both sides of the narrow street. Walking below can be perilous. In addition to water dripping from wet laundry that is hung out to dry by apartment dwellers above, water from unchecked and faulty air-conditioning units that jut out of every window—another public and environmental health risk through the potential spread of bacteria and of Legionnaires' disease—ricochets off the signs, landing on pedestrians below.[62] These foibles, however, only add to the charm and character of the "busy corner" of Hong Kong, which has managed to defy the gradual disappearance of local, family-run businesses across the rest of the city.[63] Indeed, while increased real-estate prices have led to the closure of many local, family-run businesses, Mong Kok is one area of the city where local businesses and its maze of small local stores, street hawkers, and market stalls have continued to thrive.[64]

GMK as Green Communication

Ash Amin and Nigel Thrift posit that cities can be thought of as ontographs that see people brought together—or, in some cases, separated—by infrastructures such as the positioning of streets, the design of buildings, and even the alignment of pipes. Moreover, infrastructures can act as "ontographs of power" that push, select, discriminate, bypass, watch, and engineer at a distance. Thinking of GMK as an ontograph, I will consider how the length, width, and height of GMK employ infrastructure, natural resources, the movement of inhabitants, and the flow of culture within the space to try to envision a green Mong Kok. As I turn to a critical reading of GMK, I recommend referencing the provided reproduction of Chan's plan (see figure 1).[65]

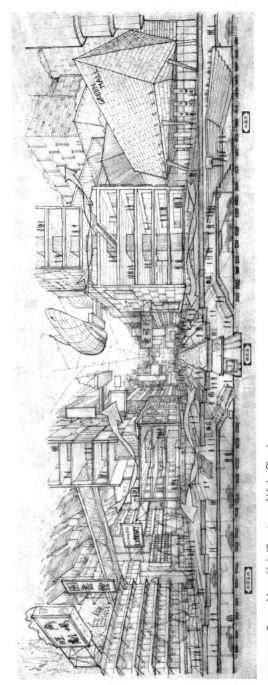

FIGURE 1. Green Mong Kok (Courtesy of Vicky Chan)

Length: The Central Spine of GMK

For length, I begin at the point that the reader's eye—or, at least, my own eye—is drawn to: the subway station in the bottom center of GMK. The further the eye moves from the close focal point of the design, the more unclear and ambiguous the features of GMK become. Like most urban areas, images of current-day Mong Kok show regular breaks in the buildings. These thoroughfares between each block enable regular access to and exit from both sides of a space. The central spine of GMK, however, appears to be continuous, with no breaks from the left or right of the space. Access to GMK, then, is governed by the subway, underground parking lots, and the air tram (which is only accessible from the right side of the space). Thus, once individuals arrive via one of these transportation modes, their movement is governed by Chan's linear design. The central spine of GMK, the area providing a sense of the length of the space, is the only area in which inhabitants are brought together and able to freely mingle. As these comments suggest, envisioning a green future is complicated by the daily realities of mobility, access, and the sheer ebb and flow of bodies moving through space.

Width: GMK as a Horizontal Space

When looking at GMK from a perspective of width, a clear divide erodes the sense of togetherness and opportunities for mingling between socioeconomic classes. When analyzing the width, I split the space into the right and left sides from GMK's central spine. In addition to the fact that the central spine offers the perfect place to divide the space into two, almost equal sides, my decision to do this stems from the fact that, on closer inspection, crossing from the right to the left side of GMK is impossible without going underground. In order to cross from one side to the other, users of the space are forced to leave through the B1 subway exit on either side of the space and reenter through the B1 exit on the opposing side of the central spine. No other pathway is clearly communicated via Chan's drawing. As an act of imagining a new and greener Mong Kok, then, GMK appears not to enhance the mobility of its citizens.

When looking across GMK, a clear divide is visible. The right side of the drawing features a single distinguishable building: a shopping mall. Containing no identifiable windows, the building is designed to hinder individuals' ability to see in or out. The only other identifiable aspect of the right side of GMK is an outdoor basketball court. Intriguingly, although outdoors and uncovered, GMK's basketball court is below street level. This feature is a nice touch, for Hong Kong has an abundance of outdoor basketball courts, which attract players of all ages and abilities. At lunchtime or after work, the city's outdoor basketball courts are often full of shirtless males playing pickup games with friends, colleagues, and strangers. Yet, the basketball court in GMK lacks sweaty, shirtless Hong Kongers. In fact, with a windowless shopping mall and a basketball court that is designed below street level, the individuals who might inhabit the right-hand side of GMK have very little chance of being seen from the opposite side of the space, or by each other. The conventions of architectural drawing preclude the kinds of novelistic detail I am thinking of here—the rich humanity that makes street life so compelling in Hong Kong—so the lack of bodies in motion within GMK invites viewers to imagine how the space might be inhabited. Nevertheless, including physical health and fun within the scope of Chan's vision is valuable to the cultural interactions one might imagine in such a space.

Two distinguishable buildings are situated on the left side of GMK: a Chinese opera theater and a library. Libraries and the Chinese opera are open to anybody, and the latter is described in Chan's plan as an aspect of Chinese pop culture. For many, though, it is notable that reading books in a library or attending the opera are often regarded as a "higher" cultural form than playing basketball or shopping in a mall. Making space for both in the design speaks to a broader commitment to not only shopping and playing but also learning and appreciating the arts.

Providing 3D printers and scanners as part of the design, GMK's information exchange is intended to provide new technologies to local merchants and students. Although the facilities and technologies provided in the information exchange may be especially useful to members of the population who do not have the economic means to own or access such equipment, the layout and design of GMK does not invite such citizens. This uninviting environment is reinforced by some of the language used within the plan. For example, the description of the information exchange as functioning "more like a think tank" than a standard library suggests a space in which experts gather to discuss, research, and affect economic policy.[66] The connotation of this language is that individuals perceived as less educated may not belong or be welcome in the left-hand side of GMK, though it is uncertain.

While buildings and spaces can be studied in a multitude of ways, Low draws attention to the underlying political and economic relations that initiate and drive their spatial production. These foci include the production of space to accommodate the desire for money and commodities, both of which are evident in GMK.[67] While both sides of GMK display a desire for cultural consumption, the difference in the types of cultural commodities that are available to individuals on different sides of the space is telling and reveals much about how the space is designed to accommodate distinct sections of Hong Kong's society. For urban eco-justice scholar Julian Agyeman, most public spaces "serve as meeting places for people who already know each other."[68] This accommodation acts as a form of "spatial injustice" by encouraging a sense of "homophilic sociability" in which individuals maintain relationships only with people who are similar to themselves.[69] For example, the allocation of physical space and natural features is prevalent on the left side of the space, but this is problematic with regard to GMK's goal of providing a democratic space. The area that is dedicated to a basketball court on the right side of GMK, for example, is replaced with a larger parking lot on the left side of the space. This decision raises a number of questions, for although evidence suggests that car ownership is increasing in Hong Kong, the costs associated with the owning and running of a car in the city are "stratospheric" and out of reach for the average Hong Konger.[70] For a future-oriented imagining of a possible eco-city, it seems a vision of urban life free from cars would be more appropriate.

The unequal use and distribution of natural elements, such as trees and light, also is visible across GMK. In contrast to the concrete shopping mall and basketball court on the right side, large trees line the left side of GMK. While research highlights that urban trees aid mental health and decrease a multitude of other health risks, they are often thought of as a luxury.[71] Although trees are viewed as "public health assets" that "make people happier," the distribution of trees in GMK leaves the city looking less environmentally sustainable, not more.[72] Here again, for a piece of imaginary green communication, it would make sense to expect to see more greenery, more foliage. Another natural resource also plays a role in GMK: sunlight. As cities grow taller, natural light has become a "precious commodity" for designers and inhabitants of urban spaces.[73] Although the façades of each building are not explicitly labeled across GMK, it is apparent that the shopping mall is made of a material that is not transparent. Conversely, structures on the left side of GMK appear to utilize materials that allow for more natural light. Numerous studies show that, like trees, exposure to natural light provides a number of health benefits while, at the same time, it increases positive moods and lowers depression.[74] The harnessing of the natural light in the built environment that is so important to health is made possible through architects' use of a particular material: glass.

In addition to being considered as versatile, incredibly strong, and aesthetically pleasing, glass has transformed modern architecture, signifying futuristic fantasies. Despite being associated with contemporary buildings, glass is one of the oldest building materials, thought to possess magical, transformational properties.[75] For example, in his essay analyzing the use of glass architecture in Macau—another of China's "Special Administrative Regions" and just a short ferry ride from Hong Kong—urban and environmental communication scholar Tim Simpson draws attention to the work of the German writer Paul Scheerbart, who was one of the leading proponents of expressionist architecture at the turn of the twentieth century. Through his fantasy writings about glass architecture, Scheerbart extolled a utopian vision in which glass replaced the brick buildings of capitalism, which he argued led to claustrophobic rooms reflecting "closed minds and closed relationships." For Scheerbart, writing in the early twentieth century, glass had a "didactic transformative potential" that could lead to architectural change and enable "wholesale cultural evolution." While brick buildings do "harm," Scheerbart declared that the choice between glass architecture and brick was not technical or economic . . . it was a moral choice.[76] A firm believer that culture is "created by architecture," for Scheerbart, a desire to improve culture begins with a change in architectural approach. This change in the "formal structure of one's surroundings . . . can actually effect a change in thinking."[77] Using glass over brick to harness natural light, then, has been imagined as capable of raising culture to "a higher level." Glass also may signify wealth today, as many top transnational corporations favor it in skylines globally, or transparency, as some corporations have begun to favor glass interior walls over opaque material choices. Given the deep architectural debates over materials, then, it is worth noting that glass is used only on one side of Chan's GMK design.[78]

No matter how one feels about glass aesthetically, environmentally, light is important for passive heating and cooling needs. As a historical symbol of knowledge, wisdom, and guidance—especially in a religious context—light also matters.[79] Therefore, it is worth noting again that light is not provided to people on the right side of GMK. This natural light that glass offers to the left side of GMK is also harnessed in another way that keeps with the intended positive environmental impacts of Chan's design but, at the same time, privileges certain sections of Hong Kong's society.

Although GMK was designed independently, without any government influence, the issue of architectural ideology means that Chan is responsible—whether consciously or unconsciously—for GMK's impact on how power dynamics exist within the space. The term *social control*—used to describe power dynamics that are present through the structuring and manipulation of space—is often associated with governments and other sovereign institutions.[80] These power dynamics are immediately evident by the fact that the lowest level of GMK is the aforementioned basketball court on the right side of the space. In addition to a clear divide within the horizontal view of GMK, I posit that vertical movement is also strictly governed. Through a number of GMK's design features, inhabitants are invited to travel through the space in a particular way. While people are naturally inclined to move vertically through GMK via stairwells, elevators, and bridges, as individuals move higher up their designated side of the space, the buildings move away from each other (perhaps to allow in more sun for the streets below). Indeed, given Hong Kong's renowned verticality, including the way its towers create a sense of shadows and darkness even in the daytime, these openings toward the top of GMK indicate a clear attempt to open up the city, to let in the light, to create a sense of openness and less geometric repetition via thousands of vertical boxes.

At the same time, however, this openness entails the reproduction of social class. For example, the lower floors of the right-hand tower are taller and wider. This design aspect suggests that more individuals are expected to be present in the right tower, with the left tower reserved for elites of a higher class. Smaller rooms and fewer floors in the left-hand tower suggest that, due to a lack of space, individuals get to ascend at a faster pace. The social inequality through the unequal allocation of space and resources is exacerbated as GMK gets higher. With the advantage of better views, reduced noise, and improved air quality, living and working on higher floors of buildings is advantageous.[81] These benefits, however, come at a financial premium, meaning that only the wealthiest Hong Kongers can afford to live, work, and exist at such high altitudes, away from the masses.[82]

Toward the top of GMK, the left side features a sky garden meant to "enhance the tie between private and public space." These "hanging oases" are meant to "drive curiosity"; however, they are fairly inaccessible and unreachable from the right side of GMK. While bridges—referred to as *sky walks*—connect various levels within each side, no bridge directly links the right and left sides of the space at the "oases" level. Although intended to transform Mong Kok into "a neighborhood without boundar[ies]," the distribution of bridges and "sky walks" might leave inhabitants wondering how to traverse the city's higher levels.[83] Still, the "hanging oases" are a nice touch, an attempt to envision a greener city.

Flow: The Infrastructure of GMK

Having now addressed the vertical and horizontal axes of the drawing, and having addressed its sense of depth, its use of materials (including trees, glass, and light), and the possible implications of its sense of space and mobility, let us now turn to the question of circulation. Referring to "the basic material and ideological structures that underlie the operation of environments," infrastructure facilitates "the flow of goods, people, or ideas and allow[s] for their exchange over space."[84] For anthropologist Brian Larkin, infrastructure comprises the architecture for circulation, supporting modern societies and generating the ambient environment of everyday life.[85] The design of the flow of culture and information, therefore, also warrants our attention.

This concept is indicated visually by four arrows, traveling right to left and back again, that sit in the center of Chan's drawing. For Chan, these four arrows represent improved airflow that will "remove stagnant pollutants" and "allow people to enjoy the neighborhood from multiple perspectives."[86] Although these arrows are labeled as presenting clean airflow, their direction and movement can also be assumed to represent the creation and flow of culture and knowledge as a part of GMK's infrastructure.

Representing the lowest level of knowledge and culture within GMK, the first arrow begins at the basketball court and moves toward the left side of the space. As the flow of culture travels across GMK, the next arrow rises from the ground as it reaches the Chinese Opera House. The third arrow appears from the direction of the library and heads back toward the right side of GMK. By this point, after traveling from the basketball court to the opera house to the library, the level of culture and knowledge present in GMK has been transformed from a form of low culture to high culture. As a result, the third culture arrow has risen away from the ground and is exactly halfway up Chan's drawing. From here, the flow and quality of culture continues to rise at it heads back to the right side of GMK. Traveling toward the shopping mall, the arrow rises sharply. This sudden rise means that the arrow misses the shopping mall and trails off into the sky.

If Chan's clean airflow indicators are recontextualized as the direction of culture flow through GMK, the arrows achieve one of Chan's intended goals but fail to meet his second goal. Success lies in the fact that the culture flow does "remove stagnant pollutants" via the cleansing of lowbrow culture that transmits from the right side of GMK and attempts to infect the left side of the space. This form of lowbrow culture is dismissed and upcycled by the left side of GMK, before being sent back to educate and improve the opposite side of the space. The fact that this repackaged culture flow misses the shopping mall and rises up and above GMK results in the failure of the second goal of Chan's airflow strategy: to "allow people to enjoy the neighborhood from multiple perspectives."[87] As I have highlighted through my analysis of GMK, even as it strives to envision a future eco-city, it also reinscribes a sense of class and order, hence reinforcing what I previously noted as "homophilic sociability." The drawing thus embodies one of the central questions of architecture and urban planning more broadly: how can we envision the future without remapping the habits and practices of the present? How can we envision the future?

(Re)Evaluating GMK

When hundreds of thousands of Hong Kongers took to the streets of their city in 2014, for seventy-nine days Hong Kong was reorganized and reconstructed into an environmentally conscious and communal space. Buoyed by this transformation, architect Vicky Chan designed a plan that he felt would ensure that the spirit of the Umbrella Revolution continued and, possibly, extended across the city. Chan claims that GMK will "maintain the core and heritage" of the district; however, many of the distinct characteristics of Mong Kok's culture are conspicuously absent from Chan's GMK.[88] Although Chan does include some neon signs in his drawing, the numbers are far less than those that are present today in Mong Kok.[89] The lack of local business, but inclusion of, for example, an opera theater suggests an upscaling plan for Mong Kok that excludes some of the lived history of everyday Hong Kongers.

Thus, I worry that rather than creating a blueprint for a democratic and sustainable district that brings inhabitants together, Chan has inadvertently created a space that reflects the deep class and socioeconomic divides of Hong Kong. More specifically, the plan separates the upper and middle classes from the city's poor and, in doing so, perhaps leaves poor Hong Kongers green with envy. Close analysis of the length, width, and height of GMK reveals that instead of promoting a sense of belongingness through the design of an environmentally friendly space that harnesses visions of connection, Chan has imagined a space that invites a sense of separation that maintains the status quo. Although the reactions of people inhabiting an imagined space are impossible to judge, for geographer Tim Edensor, despite the growth of contemporary spaces, "most of us live in recognizable worlds, distinguished by distinct material structures, distribution of objects, and institutional arrangements."[90] As an imagined future space, many of the anchors that form everyday spatial identity are present in GMK, including a shopping mall, a basketball court, trees, a library, and subway stations. As a result of containing many of the aforementioned "experiential structures of everyday life," GMK's future-based vision includes a sense of the familiar and typical—again indicating just how hard it is to envision a new way of being, our new eco-cities, without reproducing our old, environmentally disastrous cities.[91]

Indeed, with a blueprint for an imagined space, Chan faces many hurdles to overcome if his vision for GMK is to become reality—not least, GovHK. At the same time that Chan desires to create public spaces that bring citizens together, GovHK has launched a study that seeks to achieve similar results, albeit without the environmental components. While GovHK's Pilot Study on Underground Space Development in Selected Strategic Urban Areas seeks to improve pedestrian connectivity and enhance Hong Kong's living environment, the difference between the plans presented by Chan and GovHK rest in the fact that GovHK seeks to harness opportunities for utilizing underground spaces. With awards of over HK$40,000 available, Hong Kong's youth are encouraged to submit designs for utilizing "a coherent, connected, high quality, and vibrant network of underground space."[92] Although GovHK's attempts to create spaces of connection for Hong Kongers is admirable, like GMK's use of the basketball court and a windowless shopping mall, GovHK's plans appear to be pushing sections of Hong Kong society out of sight. Forcing democratic leanings underground provides yet more evidence of social control and governmentality in Hong Kong.

While Chan's ecological vision serves as a compelling reminder of the significance of imagining futurity in concrete spaces, I conclude that it would benefit from incorporating social equity and economic justice. I posit that there is more to the sustainability of a space than addressing climate change and tackling pollution, though these are vital material needs in our precarious age. Indeed, analysis of GMK demonstrates that a broader understanding of sustainability involves the consideration of sociability, economics, wealth distribution, aesthetics, opportunities, and hopes that have to be maintained and equally distributed among all inhabitants. While GMK may provide a green and environmentally friendly alternative for one of Hong Kong's most polluted districts, Chan's blueprint would be strengthened by including more elements of local history, human interaction, and the vernacular culture that provides the backbone of one of Hong Kong's most vibrant districts. Less of a divide between the left and the right side of the design might enable more interaction between inhabitants of the space.

Of course, Vicky Chan did not intend for GMK to separate inhabitants and maintain hegemonic control and governmentality. Chan is a Hong Konger who, inspired by his city's fight for democracy, desired to preserve the sense of togetherness, optimism, and inclusiveness that was present throughout the 79-day occupation of Hong Kong's streets. Having won awards for his forward-thinking environmental urban designs for spaces in China, Mexico, Saudi Arabia, South Korea, and the United States, Chan has made presentations to mayors across the globe and has taught sustainable design and urban planning to thousands of students. His work is significant, therefore, and warrants our attention.

My reading of GMK leans on the argument of architecture and urban-planning scholar Bart Verschaffel, who argues that architecture not only creates the material concrete circumstances within which a society functions, but "presents an immediate picture of a possible world . . . it proposes a normative and ideal spatial expression of that society and it represents values."[93] The potential impacts of this envisioning work, and the political conversations they foster, is why urban planning and architecture should be key parts of environmental communication. In this case, while GMK harnesses elements that are generally associated with broader economic and social equity appeal, such as an information exchange, a library, and a mass-transit system, Verschaffel's ideal notions could be more evident in GMK in ways that clearly communicate free movement, convergence, inclusiveness, and sustainability. This is no easy task, however, as imagining a green future—what Verschaffel calls "a possible world"—in one drawing amounts to a supreme act of rhetorical invention. The function of GMK is not to master this dilemma but to invite us into a conversation about what is present, what is missing, and what might be necessary to help Hong Kong move into a green future.

On the Futurity of Eco-Cities as Green Communication

When, in late 2017, Syria signed the United Nations–backed Paris Agreement on climate change, the United States became the only nation on Earth to reject the treaty. US President Donald Trump even went as far as stating that he "doesn't believe" the assessment of scientists who, throughout a 1,600-page report, highlight the grim impacts that climate change will have on every sector of American society.[94] Notably, the president chose to use China—and in fact "all of Asia"—as what Pezzullo refers to, in her introduction to this book, as the "worst-case measuring stick."[95] The truth, however, is that as Chinese architects and urban planners are producing visions similar to Chan's, China is emerging as a leader in climate policy and green leadership more broadly. The role of cities cannot be underestimated in this story.

We find ourselves at a point in history when, for the first time ever, more people in the world live in cities than in rural settings.[96] While around 55 percent of the world's population currently reside in urban areas, by 2050, that number is expected to rise to 68 percent.[97] Under its central "urban-rural integration" policy, China is the leader of what Robin Visser refers to as "the worldwide suburbanization revolution."[98] The policy, which forms part of China's previously mentioned "war against pollution," aims to move hundreds of millions of rural residents into newly constructed urban areas. In time, this mass movement will lead to the number of dwellers in China's newly developed cities almost matching the total urban population of the United States.[99]

The Party claims to be developing a staggering 285 purpose-built eco-cities.[100] One of these developments in the Chinese city of Tianjin—a groundbreaking collaboration with the Singapore government to build a "socially harmonious, environmentally friendly, and resource-conserving city"—claims to strike a balance between the competing forces of the social, economic, and environmental needs of a city.[101] These are lofty aspirations that might indicate a step in the right direction for a more environmentally friendly and just world—or they could be just more Party propaganda to support the forced migration of rural peasants into cities, where their consumer habits and political activities can be more closely monitored. While we cannot know for sure the reasons behind this explosion in Chinese "eco-cities," the issue revealed by my analysis of GMK is the underlying problem of how latent architectural and urban-planning habits and norms can infuse even imaginative renderings of the future—meaning the present impinges on the future in ways that can result in the forgetting of the very people who frequent the spaces we are trying to reimagine. It is hard to escape the pull of the present.

Hong Kong's ambiguous existence as a territory that is in a simultaneous period of decolonization (from the UK) and recolonization (by China) means that, as 2047—the date of the city's full integration with the mainland—approaches, Hong Kongers are fighting for a democratic future. While Chinese President Xi Jinping seeks to "promote green development, solve prominent environmental problems, and firmly enshrine environmental protection in the country's development path," Hong Kong's unique relationship with the mainland means that the city poses an interesting case study for a book on environmental communication and China, reminding us of the heterogeneity of the nation.[102]

Much like the designs offered for China's large-scale plans for its environmentally conscious future, analyzing Chan's vision for GMK as a form of applied intervention is worthy of our attention as it shapes choices not only in the present but also in the futures we wish to build. While closely scrutinizing plans is one way to learn from and shape environmental urban designs, studying projects that have moved beyond pencil and paper offers another method of investigation. Although such research goes beyond the scope of this chapter, by way of conclusion, I want to offer one exemplar beyond China and the United States in which citizen involvement in planning has been successful.

Once dominated by polluting coal mines and steel plants, the German city of Essen is one example of a space that has transcended its initial architectural drawings and has won awards for its transformation "from grey to green."[103] In 2017, Essen was awarded the European Commission's Green Capital Award to honor its success in embracing transformation by altering its environmental legacy. Part of Essen's success can, perhaps, be attributed to the inclusion of the city's residents in its transformation. In addition to resident-led schemes to plant trees, clean playgrounds, and offer shelter to the homeless in exchange for work as park rangers, local residents of Essen suggested many of the city's green initiatives to the government. For Markus Pajonk, lead coordinator of the city's local volunteer group, Essen Packt An, "green should be a long-term thinking of the people" and residents have to work together for the good of the urban environment.[104] Essen, then, provides evidence that a move toward the implementation of bottom-up public engagement initiatives that increase interpersonal interactions across difference can be successful.

While an increasing sense of global environmental awareness has led to a bright light being shed on sustainable architecture, eco-city attention has largely focused on the environmentally friendly aspects of buildings.[105] However, as I hope to have shown in this chapter, the form of these buildings—appearance, configuration, and shape—also bears a significant influence on the surrounding space and its inhabitants.[106] In view of this, architects, planners, and, indeed, scholars need to look beyond the obvious environmentally friendly and sustainable features of buildings and, instead, pay attention to the larger dimensions that make up the "microclimatic consequences" of urban spaces.[107] I posit that as a result of having a direct impact on the inhabitants of a space, as access to urban green space becomes an issue increasingly tied to environmental justice, these consequences must be considered in the initial conceptual and design stages of new urban environments. As cities are sites of the "(re)production of hegemonic structures," for Dickinson and Aiello, studying these spaces may lead to spatial hegemony being altered and shifted and, in doing so, "make the city more humane."[108] In short, by analyzing architectural drawings that represent visions for a greener and cleaner future, we are able to better expand our understanding of the relationship between space and the society that inhabits it.

For Agyeman, inclusive spaces act as "just environments." The issue is, however, that the design of inclusive or, at best, nonexclusive public spaces is fraught with "obstacles and challenges in engaging with difference, diversity, and cultural hetero-geneity."[109] These challenges mean that many public spaces across the globe have "reinforced divisions based upon class, race, age, or ethnicity."[110] This bias in planning has resulted in the emergence of patterns of design and management that exclude some people and reduce social and cultural diversity.[111] Plans like GMK, then, are important as they harness the potential to act as a form of protest against current conditions by driving socially just public engagement and dialogue with a view to communicating what citizens want and, importantly, what they need in the future.

As we face staggering ecological and social-justice crises globally, increasing numbers of people around the world are dreaming of a green future. To this end, Chan's vision for GMK provides a generative site for imagining not only the future of Hong Kong, but the future of environmental communication, urban communication, and interdisciplinary scholarship about space more broadly. For scholars, teachers, and activists concerned with both green communication and US-China relations, then, GMK offers a stunning site of speculation, an attempt to navigate uneven power relations, and an invitation to debate the kind of world we hope to create.

NOTES

1. Green Queen, "#UmbrellaRevolution: The Greenest Protests in the World," *Green Queen Health & Wellness Hong Kong* (blog), October 3, 2014, https://www.greenqueen.com.hk/umbrellarevolution-greenest-protests-world. For more information on the greenness of the Umbrella Revolution, see Chris Buckley and Austin Ramzy, "Hong Kong Protests Are Leaderless but Orderly," *New York Times*, September 30, 2014; Lizzie Dearden, "Hong Kong Protests: Demonstrators Clean Up and Recycle after Night of Clashes with Police," *The Independent*, September 29, 2014; Samanthi Dissanayake, "Things That Could Only Happen in a Hong Kong Protest," *BBC*, September 30, 2014, http://www.bbc.com.

2. Green Queen, "#UmbrellaRevolution."

3. Elenor Albert, "Democracy in Hong Kong," *Council on Foreign Relations*, April 5, 2019, https://www.cfr.org/backgrounder/democracy-hong-kong; Andrew Gilmore, "Hong Kong's Vehicles of Democracy: The Vernacular Monumentality of Buses During the Umbrella Revolution," *Journal of International and Intercultural Communication* (2019): https://doi.org/10.1080/17513057.2019.1646789.

4. Li Yuan, "Why Many in China Oppose Hong Kong's Protests," *New York Times*, July 1, 2019.

5. Samson Yuen and Edmund Cheng, "Hong Kong's Umbrella Protests Were More Than Just a Student Movement," *China File*, July 1, 2015, http://www.chinafile.com/reporting-opinion/features/hong-kongs-umbrella-protests-were-more-just-student-movement; William Wan, "Protesters Try to Memorialize Hong Kong's 'Umbrella Revolution' Before It Disappears," *Washington Post*, October 6, 2014. While the Umbrella Revolution inspired Chan's design, mass democracy-leaning protests are emblematic of a trend across Hong Kong. In the summer of 2019, a reported two million Hong Kong citizens took to the streets to protest against the city's chief executive Carrie Lam, and the proposed implementation of a bill that could see people in Hong Kong extradited to the mainland for questioning. For more on the 2019 protests, see Christy Leung, "Extradition Bill Not Made to Measure for Mainland China and Won't Be Abandoned, Hong Kong Leader Carrie Lam Says," *South China Morning Post*, April 1, 2019, https://www.scmp.com/news/hong-kong/politics/article/3004067/extradition-bill-not-made-measure-mainland-china-and-wont.

6. Vicky Chan, "Green Mong Kok," Avoid Obvious Architects, March 6, 2015, https://aoarchitect.us/2015/03/green-mong-kok.

7. Phaedra C. Pezzullo, "Performing Critical Interruptions: Stories, Rhetorical Invention, and the Environmental Justice Movement," *Western Journal of Communication* 65, no. 1 (Winter 2001): 1.

8. Vicky Chan, "Green Mong Kok in South Africa," Avoid Obvious Architects, March 29, 2015, https://aoarchitect.us/2015/03/green-mong-kok-in-south-africa; Chan, "Green Mong Kok."

9. To name just a few, see some of the excellent scholarship by Julian Agyeman, G. Thomas Goodnight, Jingfang Liu, and Phaedra C. Pezzullo.

10. For more on the deep class divides in Hong Kong, see Ivan Broadhead, "Hong Kong Exodus: Middle Class Leave City for Freedoms Overseas," *VOA*, November 20, 2013, https://www.voanews.com/a/hong-kong-exodus-middle-class-leave-city-freedoms-overseas/1793790.html; Te-Ping Chen and Chester Yung, "Politics, Cost of Living Push Hong Kong Residents Overseas," *Wall Street Journal*, August 21, 2013; Verna Yu, "Giving Up on Hong Kong," *New York Times*, February 18, 2015.

11. Donovan Conley, "China's Fraught Food System: Imagining Ecological Civilization in the Face of Paradoxical Modernity," in *Imagining China: Rhetorics of Nationalism in an Age of Globalization*, ed. Stephen J. Hartnett, Lisa B. Keränen, and Donovan Conley, 175–204 (East Lansing: Michigan State University Press, 2017).

12. Greg Dickinson and Giorgia Aiello, "Being Through There Matters: Materiality, Bodies, and Movement in Urban Communication Research," *International Journal of Communication* 10 (2016): 1294.

13. Doreen Massey, "Imagining Globalisation: Power-Geometries of Time-Space," in *Power-Geometries and the Politics of Space-Time: Hettner-Lectures 1998 with Doreen Massey*, ed. Michael Hoyler, vol. 2 (Heidelberg, Germany: Department of Geography, University of Heidelberg, 1999), 21. See also, Raka Shome, "Space Matters: The Power and Practice of Space," *Communication Theory* 13, no. 1 (February 2003): 40; Henri Lefebvre, *The Production of Space* (Cambridge, MA: Blackwell, 1991).

14. Dickinson and Aiello, "Being Through There Matters."

15. Phaedra C. Pezzullo and Catalina M. de Onís, "Rethinking Rhetorical Field Methods on a Precarious Planet," *Communication Monographs* 85, no. 1 (January 2, 2018): 103; Robert Cox, "Nature's 'Crisis Disciplines': Does Environmental Communication Have an Ethical Duty?," *Environmental Communication* 1, no. 1 (May 2007): 5–20.

16. Thanks to Phaedra Pezzullo for pointing out this connection. Dipesh Chakrabarty, "Postcolonial Studies and the Challenge of Climate Change," *New Literary History* 43, no. 1 (Winter 2012): 1–18.

17. I lived in Hong Kong from 2010 to 2013.

18. Setha M. Low, "Spatializing Culture: The Social Production and Social Construction of Public Space in Costa Rica," *American Ethnologist* 23, no. 4 (November 1, 1996): 861.

19. Setha M. Low, *Spatializing Culture: The Ethnography of Space and Place*, YBP Print DDA (New York: Routledge, 2017), 68.

20. Doreen B. Massey, *For Space* (London: Sage, 2015), 153.

21. Ibid., 37–38.

22. Low, *Spatializing Culture*, 40.

23. Ibid. For more on the rules of inclusion and exclusion, see Henri Lefebvre, *The Production of Space* (Oxford: Blackwell, 1991); Sally Merry, "Spatial Governmentality and the New Urban Social Order: Controlling Gender Violence through Law," *American Anthropologist* 103, no. 1 (2001): 16–29; Steven Robins, "At the Limits of Spatial Governmentality: A Message from the Tip of Africa," *Third World Quarterly* 23, no. 4 (2002): 665–689.

24. Low, *Spatializing Culture*, 34.

25. Ibid., 69.

26. Michael Peter Smith, quoted in Low, *Spatializing Culture*, 69.

27. "Hong Kong: The Facts—Population," April 2015, https://www.gov.hk/en/about/abouthk/factsheets/docs/population.pdf.

28. Wah Sang Wong and Edwin Hon-wan Chan, *Building Hong Kong: Environmental Considerations* (Hong Kong: Hong Kong University Press, 2000); "Hong Kong: The Facts—Population."

29. "The SUSDEV 21 Study," https://www.pland.gov.hk/pland_en/p_study/comp_s/susdev/ex_summary/final_eng/ch1.htm.

30. William Barron and Nils Steinbrecher, *Heading towards Sustainability? Practical Indicators of Environmental Sustainability for Hong Kong* (Hong Kong: Centre of Urban Planning & Environmental Management of the University of Hong Kong, 1999).

31. Ernest Kao, "Hong Kong's 'Ecological Deficit' Dangerously High, Say Conservationists," *South China Morning Post*, August 21, 2013.

32. Kenneth Leung Kai-cheong, "The Environmental Issues That HK Can't Afford to Overlook," *Hong Kong Economic Journal*, October 31, 2018.

33. Benjamin Haas, "Where the Wind Blows: How China's Dirty Air Becomes Hong Kong's Problem," *The Guardian*, February 16, 2017.

34. Laurel Moglen, "The Brilliant Billionaires of Hong Kong," *Forbes*, March 30, 2017; Wong and Chan, *Building Hong Kong*.

35. "The Socio-Economic Baseline," GovHK, https://www.pland.gov.hk/pland_en/p_study/comp_s/susdev/ex_summary/final_eng/ch3.htm.

36. Cannix Yau and Viola Zhou, "Any Hope for the Poor? Hong Kong Wealth Gap at Record High," *South China Morning Post*, June 9, 2017.

37. For information regarding Hong Kong's wealth disparity, see "Hong Kong's Appalling Wealth Gap Is a Burning Fuse for Revolution," *South China Morning Post*, October 13, 2016; "Local Wealth Inequality Worsens as Richest Earn 29 Times More Than Poorest," *Oxfam Hong Kong*, October 11, 2016, http://www.oxfam.org.hk; Yau and Zhou, "Any Hope for the Poor?"

38. "Hong Kong: The Facts—Population."

39. "Hong Kong Flats under HK$10,000 a Month Almost 'Impossible' to Find," *South China Morning Post*, October 10, 2017; Beckie Strum, "Hong Kong, Sydney Are World's Most Unaffordable Cities," *Mansion Global*, January 23, 2018.

40. Tongjin Yang, Weiping Sun, Juke Liu, Dingjun Wang, and Gang Zeng, "Ecological Crisis, Eco-civilization, and Eco-cities," in *China's Eco-City Construction*, 1st ed., ed. Jingyuan Li and Tongjin Yang (Berlin: Springer, 2016), 3–48, 8, 28.

41. Li and Yang, *China's Eco-City Construction*, 5.

42. Huifeng Li and Martin de Jong, "Citizen Participation in China's Eco-City Development. Will 'New-Type Urbanization' Generate a Breakthrough in Realizing It?," *Journal of Cleaner Production* 162 (2017): 1086; Haiyan Lu, Martin de Jong, and Ernst ten Heuvelhof, "Explaining the Variety in Smart Eco City Development in China: What Policy Network Theory Can Teach Us about Overcoming Barriers in Implementation?," *Journal of Cleaner Production* 196 (September 2018): 135–149; Jingfang Liu and G. Thomas Goodnight, "China and the United States in a Time of Global Environmental Crisis," *Communication and Critical/Cultural Studies* 5, no. 4 (December 2008): 416–421; Phaedra C. Pezzullo, "'There Is No Planet B': Questions during a Power Shift," *Communication and Critical/Cultural Studies* 10, no. 2–3 (September 2013): 301–305.

43. Michael Greenstone, "Four Years after Declaring War on Pollution, China Is Winning," *New York Times*, March 12, 2018.

44. Sha Song, "Here's How China Is Going Green," *World Economic Forum*, April 26, 2018, https://www.weforum.org.

45. Greenstone, "Four Years after Declaring War on Pollution."

46. See Robin Visser, "The Chinese Eco-City and Suburbanization Planning: Case Studies of Tongzhou, Lingang, and Dujiangyan," in *Ghost Protocol: Development and Displacement in Global China*, ed. Carlos Rojas and Ralph Litzinger (Durham, NC: Duke University Press, 2016), 39–40; Lu, de Jong, and Heuvelhof, "Explaining the Variety in Smart Eco City Development in China."

47. Sarah Jacobs, "12 Eerie Photos of Enormous Chinese Cities Completely Empty of People," *Business Insider*, October 3, 2017.

48. Jane Cai, "How China's Rush to Urbanise Has Created a Slew of Ghost Towns," *South China Morning Post*, March 8, 2017.

49. Lu, de Jong, and Heuvelhof, "Explaining the Variety in Smart Eco City Development," 138, 146.

50. Martin de Jong, Dong Wang, and Chang Yu, "Exploring the Relevance of the Eco-City Concept in China: The Case of Shenzhen Sino-Dutch Low Carbon City," *Journal of Urban*

Technology 20, no. 1 (January 2013): 95–113, 97.

51. Li and Yang, "China's Eco-City Construction," 8–9.

52. About Vicky Chan," Archinect.com, accessed December 5, 2018, https://archinect.com/vickychanarchitect.

53. Chan, "Green Mong Kok in South Africa."

54. Vicky Chan (founder, Avoid Obvious Architects), personal communication with the author, June 2015, Hong Kong.

55. Ibid.

56. Chan, "Green Mong Kok."

57. Ng Kang-chung, "Air Pollution Worst in 18 Years in Shopping Areas," *South China Morning Post*, October 26, 2016.

58. "Touring the Mongkok Ladies Market," TripSavvy, https://www.tripsavvy.com/mongkok-ladies-market-tour-1536032.

59. Ernest Kao, "Hong Kong Air Pollution Still Far Exceeds WHO Levels and Worsening, Concern Group Finds," *South China Morning Post*, July 13, 2016; "Mong Kok Affected by Pollution from Thousands of Restaurant Kitchens, Study Finds," *South China Morning Post*, August 16, 2015.

60. "Air Pollution Worst in 18 Years"; "Mong Kok Affected by Pollution from Thousands of Restaurant Kitchens."

61. "Air Pollution Worst in 18 Years"; Friends of the Earth (HK), "The Solution to Hong Kong's Serious Roadside Air Pollution Problem Is Easy," July 1, 2017, https://www.hongkongfp.com/2017/01/07/the-solution-to-hong-kongs-serious-roadside-air-pollution-problem-is-easy.

62. Sarah Karacs, "Dripping Hong Kong Air-Conditioning Units Spew Unresolved Nuisances," *South China Morning Post*, August 6, 2015.

63. For more information on the disappearance of local Hong Kong businesses, see Eric Cheung, "Family Firms Priced Out as Chain Stores Dominate up to 96% of Mall Space—Survey," *Hong Kong Free Press*, July 23, 2015.

64. Jun Concepcion, "Soaring Rents in Prime Locations Push Out Local Shops and Restaurants," *South China Morning Post*, March 18, 2018; Lizzie Meager, "Hong Kong's Excessive Retail Rental Prices Are Damaging Local Trade," *European CEO*, September 25, 2015, https://www.europeanceo.com; Bettina Wassener and Mary Hui, "Hong Kong Rents Push Out Mom and Pop Stores," *New York Times*, July 3, 2013; Derek Guthrie, "Hong Kong Shopping Guide: The Markets of Mong Kok," *The Guardian*, October 31, 2016.

65. For publication purposes, the version of GMK included in this volume (figure 1) has been altered to remove text detailing a number of buildings and other features. A drawing

of GMK with text can be found at https://aoarchitect.us/wp-content/uploads/2015/03/green-mongkok-printout-2.jpg.

66. Peter T. Leeson, Matt E. Ryan, and Claudia R. Williamson, "Think Tanks," *Journal of Comparative Economics* 40, no. 1 (February 2012): 62–77.

67. David Harvey, *Spaces of Global Capitalism: Towards a Theory of Uneven Geographical Development* (New York: Verso, 2006); Sharon Zukin, *Landscapes of Power: From Detroit to Disney World* (Berkeley: University of California Press, 1991).

68. Julian Agyeman, "Interculturally Inclusive Spaces as Just Environments," Social Science Research Council, June 27, 2017, https://items.ssrc.org/just-environments/interculturally-inclusive-spaces-as-just-environments.

69. Monique Pinçon-Charlot and Michel Pinçon, "Social Power and Power over Space: How the Bourgeoisie Reproduces Itself in the City," *International Journal of Urban and Regional Research* 42, no. 1 (January 1, 2018): 115–125.

70. Cheung Chi Fai, "Hong Kong Needs More Roads to Cope with Car Growth," *South China Morning Post*, March 18, 2018.

71. Teresa Mathew, "Street Trees Are Public Health Assets," CityLab, October 3, 2017, https://www.citylab.com/environment/2017/10/how-should-we-fund-urban-forestry/541833.

72. Ibid.

73. Lex Berko, "The Precious Commodity of Urban Sunlight," CityLab, April 26, 2014, http://www.theatlanticcities.com/design/2014/04/precious-commodity-urban-sunlight-best-cityreads-week/8963.

74. For information about the health and well-being properties of sunlight, see Berko, "The Precious Commodity"; Stephen J. Genuis, "Keeping Your Sunny Side Up," *Canadian Family Physician* 52, no. 4 (April 10, 2006): 422–423; Rosalie Chan, "Why the Summer Solstice May Be the Happiest Day of the Year," *Time*, June 19, 2016.

75. Tim Simpson, "Scintillant Cities: Glass Architecture, Finance Capital, and the Fictions of Macau's Enclave Urbanism," *Theory, Culture & Society* 30, no. 7–8 (December 2013): 343–371.

76. Ibid., 349.

77. Rosemarie Haag Bletter, "Paul Scheerbart's Architectural Fantasies," *Journal of the Society of Architectural Historians* 34, no. 2 (May 1975): 87.

78. Karrie Jacobs, "Rediscovering Paul Scheerbart's Glass-Inspired Modernism," *Architect* magazine, February 5, 2015.

79. Franz Rosenthal "Knowledge Is Light," in *Knowledge Triumphant* (Leiden, Netherlands: Brill, 2006), 155–193; "The Role of Light in Philosophy (2008)," http://www.the-philosopher.co.uk/2008/05/the-role-of-light-in-philosophy-2008.html.

80. Michael Foucault, *Security, Territory, Population—Lectures at the College de France, 1977–78* (London: Palgrave Macmillan); Low, *Spatializing Culture.*

81. Cheuk Fan Ng, "Living and Working in Tall Buildings: Satisfaction and Perceived Benefits and Concerns of Occupants," Frontiers in Built Environment, December 1, 2017, https://doi.org/10.3389/fbuil.2017.00070.

82. C.Y. Jim and Wendy Chen, "Value of Scenic Views: Hedonic Assessment of Private Housing in Hong Kong," *Landscape and Urban Planning* 91, no. 4 (July 30, 2009): 226–234; Raymond Tse, "Estimating Neighbourhood Effects in House Prices: Towards a New Hedonic Model Approach," *Urban Studies* 39, no. 7 (June 2002): 1165–1180.

83. Chan, "Green Mong Kok."

84. Low, *Spatializing Culture*, 73; see also Brian Larkin, "The Politics and Poetics of Infrastructure," *Annual Review of Anthropology* 42 (October 21, 2013): 328.

85. Larkin, "The Politics and Poetics of Infrastructure."

86. Chan, "Green Mong Kok."

87. Ibid.

88. Ibid.

89. An image of the currently configured space of Mong Kok can be found at https://www.theguardian.com/travel/2016/oct/31/hong-kong-shopping-guide-mong-kok-street-markets#img-1.

90. Tim Edensor, *National Identity, Popular Culture and Everyday Life* (New York: Berg, 2002), 51.

91. Roger Silverstone, ed., *Visions of Suburbia* (New York: Routledge, 1997), 10; Edensor, *National Identity.*

92. "Pilot Study on Underground Space Development in Selected Strategic Urban Areas," http://www.urbanunderground.gov.hk./main.php.

93. Bart Verschaffel, "Semi-Public Spaces: The Spatial Logic of Institutions," in *Does Truth Matter?*, ed. Raf Geenens and Ronald Tinnevelt (Dordrecht: Springer Netherlands, 2009), 133–146.

94. Robinson Meyer, "Syria Is Joining the Paris Agreement. Now What?," *The Atlantic*, November 8, 2017.

95. Caitlin Oprysko, "'I Don't Believe It': Trump Dismisses Grim Government Report on Climate Change," *Politico*, November 26, 2018.

96. "Two-Thirds of World Population Will Live in Cities by 2050, Says UN," *The Guardian*, May 16, 2018.

97. "The World's Cities in 2016," United Nations, https://www.un.org/en/development/desa/population/publications/pdf/urbanization/the_worlds_cities_in_2016_data_booklet.pdf;

"Two-Thirds of World Population."

98. Visser, *The Chinese Eco-City*, 37.

99. For more on China's "war on pollution," see Michael Greenstone, "Four Years after Declaring War on Pollution, China Is Winning," *New York Times*, March 12, 2018. For more on US/China city dwellers, see Ian Johnson, "China's Great Uprooting: Moving 250 Million into Cities," *New York Times*, June 15, 2013.

100. Wade Shepard, "No Joke: China Is Building 285 Eco-Cities, Here's Why," *Forbes*, September 1, 2017.

101. "Tianjin Eco-City: A Model for Sustainable Development," Sino-Singapore, August 25, 2017, https://www.tianjinecocity.gov.sg/bg_masterplan.htm.

102. Chris Zhang, "Can Red China Really Be the World's New Green Leader?," *The Diplomat*, November 15, 2017.

103. Simone d'Antonio, "Essen's Award-Winning Blueprint for Greening the Post-Industrial City," Our World, February 22, 2017, https://ourworld.unu.edu/en/essens-award-winning-blueprint-for-greening-the-postindustrial-city; "Welcome to Essen, European Green Capital 2017," European Commission: Magazine Environment for Europeans, January 19, 2017, https://ec.europa.eu/environment/efe/themes/welcome-essen-european-green-capital-2017_en.

104. d'Antonio, "Essen's Award-Winning Blueprint."

105. Ilaria Giovagnorio and Giovanni Chiri, "The Environmental Dimension of Urban Design: A Point of View," in *Sustainable Urbanization*, ed. Mustafa Ergen (IntechOpen, 2016), 37–60.

106. Ibid. For more on urban form and health, see UCL Institute for Environmental Design and Engineering, "The Built Environment," https://www.ucl.ac.uk/bartlett/environmental-design/research/research-themes-methods-and-impact/built-environment-building-performance-and-processes; Harvard Center for Green Buildings and Cities, http://harvardcgbc.org.

107. Giovagnorio and Chiri, "The Environmental Dimension of Urban Design," 40.

108. Dickinson and Aiello, "Being Through There Matters," 1305.

109. Agyeman, "Interculturally Inclusive Spaces."

110. Anastasia Loukaitou-Sideris, "Children's Common Grounds: A Study of Intergroup Relations among Children in Public Settings," *Journal of the American Planning Association* 69, no. 2 (June 30, 2003): 131.

111. Setha M. Low, Dana Taplin, and Suzanne Scheld, *Rethinking Urban Parks: Public Space and Cultural Diversity*, 1st ed. (Austin: University of Texas Press, 2005), 1.

Conclusion

The study of communication is an open-ended project; yet, our natural resources are finite. In early human history, travels were dangerous, innovations slow to disseminate, and distances vast; modern communication spreads messages more rapidly, enabling media, commerce, and information to take great leaps forward. The web of our communication, therefore, is both expanding and accelerating; yet, the ecological arrangements of the planet remain constant, creating an inherent tension between communication capabilities and environmental limits.

In response to this dilemma, environmental communication has developed into a flourishing academic field in the United States and globally. In China, environmental communication is emerging rapidly, even though it has developed as a new academic field over just the past decade or so, as a younger generation of scholars have begun to respond to the opportunities and consequences of rapid modernization. Still, Chinese scholarship appears emergent, as this new cohort defines key terms, borrows and debates Western theories, incorporates narratives and concepts from our own national traditions, explores research topics, and co-constructs the field with our national and international colleagues. Further, the practices of environmental communication have gone through rapid development in the past thirty years, as NGOs, NPOs, ad hoc groups, research institutes, social enterprises, and allied organizations have sprung up all around China. Environmental communication in China has advanced so quickly that scholars, policymakers, and environmentalists alike must race to keep pace with local, national, and global change.

As the chapters herein indicate, nations exhibit differing environmental histories and practices. For example, in comparison to the cacophony of political practices in the United States, environmental communication in China tends to involve multiple parties working mostly collaboratively toward the same goal of a better environment (and society) under common cause. This is not to say that there are no tensions, fights, struggles, contestations, and confrontations, for drama and conflict in the face of hazard, risk, and danger are aspects of teaching, studying, and practicing environmentalism in China. Yet, at this time, compared to the contemporary United States, China has more unity in the creative and generative processes identified as "green public culture." It goes without saying that this unity is led by the Party, which puts social stability above all other social concerns. Still, in the case of this Party-led unity of purpose, it seems evident that having a central authority with a clear long-term vision contributes to a sense of cohesion and purpose surrounding environmental policies and practices.

Beyond the themes of care, crisis, and futurity, this book has moved us along to identify the major parties or stakeholders in environmental or green communication in China. From the preceding chapters and drawing upon the emergent body of literature on green communication in China, I identify four tracks that have played an active role in the formation of green public culture in China: the state policy and action; the Chinese government; environmental journalism and mass media; environmental nongovernmental organizations (ENGOs); and public opinion and participation. After doing so, I identify implications and future areas for research. In this way, I offer readers both a summary of the arguments made in this book and a glimpse into the future of environmental communication in China and beyond. As should be clear, I assume that the future of US-China relations hinges in part on how our two nations can learn to work together to address and solve our most pressing environmental issues.

The State Policy and Action

"From Green Peacock to Blue Sky: How ENGOs Foster Care through New Media in Recent China," which I cowrote with Jian Lu, identifies a series of important environmental laws recently established in China. Different from US environmental struggles today, the Chinese state is a key agency, stakeholder, and participant in China's cause of environmental protection. In fact, some observers argue that environmental communication in China primarily involves government-led activities. Caring for environmental protection long has been pushed forward by the Chinese government, which influences the direction and content of the practices. Presently, China is going through a transformative period that focuses on economic development and crafting a harmonious society. A central task or dilemma the Chinese government faces is balancing conflicting goals, including sustaining economic development while improving environmental protection. For as much as 20 percent of the Earth's population, this balancing act is awesomely complicated. Still, despite the obstacles of managing immense space, more than a billion people, and the world's fastest-growing economy, the central government has encouraged environmental protection by trying to reduce air and water pollution, addressing the causes and consequences of climate change, and responding to innumerable environmental crises.

Since the 1970s, the Chinese central government has made environmental protection a national goal and policy priority, with sustainable development a key strategy. Perhaps the first nationally led idea and effort of environmental protection was initiated by the farsighted Premier Zhou Enlai, as he sent a delegation to attend the first United Nations Conference on Protecting the Human Environment in Stockholm, Sweden, in 1972. In 1984, fourteen years after the United States established a national Environmental Protection Agency (EPA), China established the State Environmental Protection Administration (SEPA). Since then, the government has devoted many efforts to developing environmental policies and initiatives. Perhaps most importantly, the Environmental Impact Assessment (EIA) law was created in 2002 as the basic legislative framework for public participation in the planning stage of development projects.[1] Then the Environmental Protection Bureaus (EPBs) at different administrative levels were established to enforce environmental regulations and various modes for public participation (phone hotlines, questionnaires, public hearings). In 2008, the Ministry of Environmental Protection (MEP) was created.[2] Since then, many national strategies have addressed environmental policy, such as climate change and renewable energy.[3] What is interesting about these early developments in environmental policy in China is how they were driven in large part by initiatives coming from the United States and the United Nations; at this stage, therefore, China's environmental policies appeared to mainly be playing catch-up.

More recently, however, China has started to become an international environmental leader. China's President Xi Jinping has prioritized two key green discourses: "ecological civilization" and the "Two Mountains" theory (which is the core of the former), emphasizing the extreme importance of balancing environmental protection with economic development. In the past ten years, the content of both has been further developed and enriched, and their importance has been recognized and strengthened by the Chinese Communist Party and the Chinese government.[4] Additionally, the report "Lucid Waters and Lush Mountains Are Invaluable Assets: China's Ecological Civilization Strategy and Action" was issued by the United Nations Environment Programme, indicating that the two green discourses are not only valuable to China, but can be considered templates for pursuing a greener future globally.[5] As this moment indicates, China is no longer simply playing catch-up to the West, instead now pursuing a leadership role on a global scale.

Indeed, while China is determined to solve its many domestic environmental problems, it also has tried to play an increasingly active role in global environmental protection. For instance, the Chinese government participated in the United Nations Climate Change Conference in Copenhagen in December 2009, making a commitment to reduce carbon emissions to the world.[6] Again, in the most recent COP24 (the 24th Conference of the Parties to the United Nations Framework Convention on Climate Change) in December 2018, China was determined to play a greater and more important role in global climate governance by building a collective leadership system, increasing domestic green, low-carbon development, and working with other key members of the international community. In this role, China represented the demands of smaller developing countries in acquiring greater financial and technological support and worked closely with the European Union (EU) to push for having a common reporting and accounting system that can be applied to all countries' emissions-reduction efforts.[7]

As China extends its environmental governance globally, some scholars have argued that it makes sense for the nation's environmental policy to be considered an extension of soft power in the international arena. For example, as early as

2008, the documentary *Nourished by the Same River* (*tongyin yijiangshui*) argued for China's role as an environmental leader. As the first documentary reflecting on the environmental consequences of rapid development in the five countries in the Lancang-Mekong river subregion in South Asia, it pictured how "China attempts to reassure its neighbors by claiming that its growing regional importance is a win-win situation for all."[8] Thus, the film seeks both to reflect and amplify China's soft power in the ASEAN (Association of Southeast Asian Nations) region.

Likewise, in this book, Junyi Lv and G. Thomas Goodnight reflect in "Material Cultural Diplomacies of the Anthropocene: An Analysis of the Belt and Road Initiative between China and Oceania" on one of China's latest major development initiatives, One Belt, One Road (OBOR). The OBOR focuses on infrastructure-building investments in countries along the land routes (the belt) and the sea route (the road) in European, Asian, and African countries. Their chapter demonstrates another case in which China is considered to be overtaking the United States by adopting a leadership position in mitigating global climate change (with neighboring small island countries). By focusing on three small islands along the "Road" that are vulnerable to climate change, they argue that how China develops relationships with these countries under the OBOR demonstrates a new form of soft power—and possibly develops a transformative Anthropocene cultural diplomacy. At the same time in reality, however, communities in Asia and Africa are beginning to push back against the OBOR, and to set up their own initiatives, such as the Freedom Corridor and Partnership for Quality Infrastructure (PQI), to balance the influence of the OBOR from China. Heated discussion and debates have also appeared in the West questioning the true nature, feasibility, and global impact of the OBOR.[9] In future research, we as scholars of environmental communication should heed to such different sides of the stories and assess them with a critical posture.

Future scholars might want to address the OBOR not only as a case study of massive economic development on a regional scale, but also as a remarkably rich moment for considering the intersections of economic development, political relations, and environmental policy. Particularly as regards the future of US-China relations, OBOR indicates to the United States China's efforts to wrest global leadership from it. It will be interesting to watch how China may strive for this goal, noting whether and how its pledges about environmentally friendly forms of development are eventually implemented.

Environmental Journalism and Mass Media

In addition to the policy and legal efforts of the Chinese government, environmental journalists have acted as pioneers in environmental communication. Here, I will highlight the most significant milestones. First, in 1984, *China Environmental News* was born, marking the beginning of professional environmental-themed media in China. By 1986, the China Environmental Journalists Association was established in Beijing and included ten newspapers as the main body. A number of mainstream newspapers in the association, such as *People's Daily, Guangming Daily, and China Youth Daily,* began running environmental stories on a regular basis. At this stage, the main content of environmental reporting was Party-produced propaganda celebrating the government's major environmental protection activities, including especially its efforts to reduce the "three wastes" (wastewater, waste gas, and waste residue).[10] Numerous impressive environmental reporting appeared then, such as the famous "Woodcutter, wake up!" (Famuzhe xinglai!) article by journalist Gang Xu from *People's Daily*. This story discussed the harm of logging to human beings during a time when China's economic construction was the primary task of the society, and the environmental awareness of the whole society was quite weak.[11] In addition, many newspapers, periodicals, and radio and television stations opened columns and special editions on environmental protection. For instance, the famous *Human and Nature* and *Animal World* programs were produced by CCTV and quickly became known by every household.[12] Throughout this period, such early environmental reporting and programs sought to bring environmental issues to the public while still supporting the Party's efforts to focus on economic development.

In the 1990s, environmental reporting took a more aggressive stance exposing various environmental issues, growing rapidly in terms of the depth, intensity, and breadth of the coverage; journalists also began to cover many regional environmental problems, rather than focusing primarily on the central regions. The field of environmental news grew rapidly with diverse forms of reporting in newspapers, television, and radio. In 1993, the "Century Journey of Environmental Protection in China" was launched by the Environmental Information Committee of the National People's Congress, the Ministry of Propaganda, the Ministry of Radio and Television, and the State Environmental Bureau. This Party-led initiative promoted the rapid development of environmental news reporting in China and sent a key message to activists as well: environmental advocacy was henceforward welcomed by the Party. Echoing this, twenty-eight mass media in Beijing carried out specialized environmental reporting, creating a wave of new consciousness about environmental issues. These newly empowered reporters exposed issues such as the pollution of the River Huai, which experienced several severe industrial accidents; the killing of the Tibetan antelope; the destruction of China's ancient forests; and the previously noted birth of China's environmental NGOs (ENGOs).[13]

With the rapid economic development of the twenty-first century rocketing China toward the world's most intense burst of construction, so environmental accidents have multiplied rapidly, causing tremendous concern both domestically and abroad. For example, in the past decade China's media consumers have seen a steady stream of stories about food safety, resource shortages, the consequences of overconsumption, controversies over public health and disease, and others.[14] Environmental journalists exposed many sensitive topics such as the notorious "cancer village" and the "Blood Lead Village."[15] In addition, environmental investigative journalists, who emerged in China in the late '90s, exposed wrongdoing in the process of dam-building on many rivers, such as the Yang Liu Lake, the Three Gorges, and the Nu River; in each case, brave journalists exposed patterns of malfeasance and corruption impacting the local environment and local populations.[16] As famous environmental events occurred—such as the Anti–Nu River Dam Building campaign since 2004, the Old Summer Palace Anti-Seepage Accident in 2005, the Taihu Lake Cyanobacteria Pollution in Wuxi in 2007, and the PX Project in Xiamen in 2007—environmental journalism publicized these events, generating attention and increased public participation in response.[17] It was the environmental journalists who served as a critical voice for Chinese society; they made voices from the bottom of society be heard, and encouraged exchange of diverse social thoughts and cultural values on the relation between nature and the human. In addition, this stage in China's evolving environmental journalism shows an interesting contradictory pattern of development—environmental journalism both continued to serve as the mouthpiece of the central government by fulfilling its role of public opinion supervision on environmental issues, and began to challenge the interests of the state by criticizing individual businesses and governments at the local level.[18]

Since the 1990s, China's mass media also produced many environmental documentaries. For example, in 2004, CCTV premiered *Forest China* (*Sen Lin Zhi Ge*), an eleven-hour eco-documentary covering the major forest areas in China, including the Northeast Forest, Qinling Mountains, Taklimakan Desert, Southeast Tibet Forest Area, Hainan Tropical Rainforest, South China Sea Mangrove, and so on. *Forest China* shows the magic and beauty of the forest with exquisite pictures, as well as the harmonious symbiosis of humans, animals, and forests.[19] Critics might point to the documentary as not so much reporting as propaganda, yet like National Geographic episodes in America, the film works to cultivate a sense of love—what Phaedra Pezzullo has called an ethic of "care" in the introduction—for the natural world, hence inculcating values amenable to environmental activism. On the other hand, recent years have also seen the release of some sensational environmental documentaries, such as *Plastic Kingdom* and *Beijing Besieged by Waste*, produced by journalist Jiuliang Wang in 2015, and the widely known *Under the Dome*, produced by Jing Chai (also a journalist) in 2016. These independent films enjoyed great success, gained national and global attention, and depicted in harrowing terms the consequences of unchecked human development.

Of course, mass-media productions of environmentally themed work is not limited to nonfiction films made by Chinese directors. For example, in "The STEMing of Cinematic China: An Ecocritical Analysis of Resource Politics in Chinese and American Coproductions," Pietari Kääpä provides an "etic" eco-critical analysis of two Hollywood-style films—*Skyscraper* and *The Meg*—to illustrate how their Sino-US coproduction (culturally and politically) offers a new popular image of China, which is shown using science and technology to approach natural resources and energy. Likewise in their chapter, we saw Lv and Goodnight argue that media depictions of China's managing of OBOR indicate China's attempts to merge its political ambitions with a sense of soft power based, in part, on the nation's environmental policies. As these examples indicate, China is emerging quickly in the areas of international soft power, while the nation has evolved rapidly as a hotbed for environmental journalism and mass-media production. The key question for future environmental communication scholarship to ask, then, is: what kind of role can environmental reporters and mass media (continue to) play in terms of changing China's environmental and political landscape?

Environmental Nongovernmental Organizations (ENGOs)

Alongside China's environmental journalism, scholars have highlighted the role of environmental nongovernmental organizations (ENGOs) in leading China's green movement, analyzing these ENGOs as a countermovement and as playing a super-visory function. [20] Indeed, the birth of Friends of Nature in 1993 is widely considered as marking the emergence of China's third notable voice in environmental public culture, and as indicating the beginning of China's green social movement. As a major movement player, ENGOs have enjoyed vibrant growth in the last two and half decades, with numbers growing from 3,539 in 2008 to 7,881 by the end of 2012.[21] As in the United States as well, these Chinese ENGOs represent a dizzying range of organizational structures, goals, communicative tendencies, and membership bases, placing them among the forefront of emerging democratic practices in China.

While appearing to specialize by specific interest and regional location, when considered collectively these ENGOs articulate a wide variety of issues including but not limited to recycling, animal rights, water pollution, air pollution, biodiversity, and environmental rights. They have initiated an overall vibrant, growing green movement with unique styles. The results have been mixed successes and failures. While Chinese ENGOs have not mobilized the wider Chinese public to a large extent to engage in environmental movement, they have achieved some signature results. Three examples of well-known successful campaigns ENGOs led in China include the Saving the Tibetan Antelope campaign in 1999, which successfully reversed the trend of antelope poaching; Protecting the Old Summer Palace campaign, which urged the Environmental Protection Bureau (EPB) to hold the first administrative hearing under the EIA process in 2004; and the successful ending of the Anti–Nu River Dam campaign that lasted from 2003 to 2017. These three environmental victories stand in recent Chinese political life as potential harbingers of things to come, as Chinese citizens, activists, journalists, and government officials worked together for the common good of the nation.

In their first ten years of development (1994–2004), Chinese ENGOs focused on environmental education, animal protection, and promoting public supervision. Many of these "older" ENGOs were initiated by a generation of charismatic leaders, including Zhinong Xi, who initiated Wild China Film, and Yongchen Wang, the former CCTV reporter who established Green Earth Volunteers. Moreover, many of these ENGOs have created some of China's first independent media products. For instance, since 2005, Yongchen Wang began to host a series of "Environmental Journalist Salons" and, since 2008, to publish annual reports of Chinese environmental journalists' investigations. Zhinong Xi, China's first wildlife photographer, also has led his Wild China Film photographer team to produce China's first notable and highly circulated wildlife photographs of Tibetan antelopes, Yunnan snub-nosed monkeys, green peacocks, and other endangered species.

In the second decade of Chinese ENGO development, beginning around 2004, ENGOs featured more rational and scientific styles in terms of operations and targeted public advocacy, often marshaling the efficient use of information and communications technologies (ICTs). Consider, for example, the Institute of Public and Environmental Affairs (IPE), initiated by Ma Jun and based in Beijing. IPE focuses on the rich uses of ICT to publicize the information disclosure of governmental agencies and enterprises to promote transparency of data and to improve environmental governance. It created two unique and notable databases on the Internet with mobile applications or apps: China Water Pollution Map and Blue Sky app (China Air Pollution app). Both play a vital role in supervising the government and enterprises through new media-based information disclosure. The use of ICTs for environmental activism is an especially rich area for future scholarship, as it demonstrates the interface of advocacy, technology, and public participation, for as such technologies become operationalized in user-friendly apps, it will be interesting to watch whether and how a new generation of news and social media users can become energized activists by simply accessing an icon.

In fact, the third generation of Chinese ENGOs has mushroomed since roughly 2014. I argue that they operate with an even more sophisticated style. Some have begun to take the new organizational format of "social enterprise" to merge nonprofit and business models.[22] These newer ENGOs tend to seek cooperation with business and government agencies, and are particularly adept at engaging the social media and pan-media opportunities for their cause. Future scholarship on these third-generation Chinese ENGOs will want to pay attention, in particular, to how WeChat and Weibo play different yet crucial roles in contemporary ENGO communication.

Throughout the development of ENGOs in the past twenty-five years, new media have been used proactively as a crucial tool for advocacy work, from pushing for information disclosure to publicizing, recruiting, and organizing mobilization and campaigns online and offline.[23] New media activism of Chinese ENGOs is a promising field that deserves much attention. It is even more fascinating to compare how Chinese and US environmental activists engage in new media activism across different platforms, under different conditions and strategies, and to achieve different results. In terms of US-China relations, it will be equally fascinating to watch as Chinese and US NGOs begin to cooperate more, perhaps even developing the first generation of transnational environmental apps.

Not only are Chinese ENGOs playing a crucial role in mitigating environmental crises within China, the recent years have also seen them increasingly active in assisting with global climate governance. Take the recent COP24 as an example.

On December 5th in the China Pavilion of COP24, a side meeting, "Global Climate Governance and NGO Contribution," was cosponsored by the China Association for the Promotion of International Cooperation among Civil Organizations and a couple of other organizations. At this meeting, multiple parties shared the current situation of global climate governance and the successful experience of ENGO participation. Participants shared how China's ENGOs have cooperated with the Chinese government and enterprises to play a positive role in addressing climate change.[24] This meeting is an example of how studies of Chinese ENGOs can be examined from different perspectives aside from their new media activism, including organizing coalitions, international public participation, and discourses about climate change. The trick with this new wave of ENGO work, of course, is to watch how locally or regionally based movements attempt to link up with regional or national or even international movements under different sociopolitical circumstances. As has been true of environmental activism for decades, then, the question remains: how can even these sophisticated, third-generation Chinese ENGOs of different generations map their local concerns onto the issues of other national and international advocates? How can "care" transcend the nation-state to become a truly cosmopolitan movement?

Public Opinion and Participation

Although Chinese ENGOs have grown in number and diversity, they have done so under a stringent sociopolitical situation in China.[25] As such, ENGOs have not been completely successful in changing public opinion in ways that can keep pace with the growing environmental degradation in China. In the Chinese context, then, ENGOs have struggled to make a strong public impact due largely to the constrained context of public political action. The US context illustrates the opposite problem: while ENGOs have blossomed, and while the public sphere is open and thriving, environmental activists have had trouble cutting through the saturated media to make as big an impact on public opinion and government policies as desired. In the former context, ENGOs suffer from a deficit of support and communication; in the latter, they suffer from a surplus of noise.

Given these complex and often contrasting situations, cross-cultural analyses of public opinion about and participation in environmental issues, in both the United States and China, are needed. Thus, in their chapter "Comparing Chinese and American Public Opinion about Climate Change," Binbin Wang and Qinnan (Sharon) Zhou offer an important example of how to study different public attitudes toward climate change in China and the United States, hoping to shed light on why the two countries have such different approaches. Wang and Zhou found that while Americans are more anxious about climate change than the Chinese, the Chinese have more positive feelings about the impact of and desire for climate-change education. Across both countries, the common trends indicate that the majority of publics in China and America support climate-change actions and sustainable development for a collective future.

Public opinion, of course, is shaped by experience. To consider another significant voice in China's green public culture, I now will focus on the role of green protests and ecotourism as forms of environmental communication that exceed government, journalists, and ENGOs. In contrast to organized ENGO activism, self-organized mass protests by local citizens have recently grown quickly in China to challenge developed industrial projects. According to government statistics, environmental "mass incidents" or protests grew by an average of 29 percent a year between 1996 and 2011. Since 2005, the MEP recorded a total of 927 environmental incidents; among them, 72 were considered major events. The year of 2011 alone witnessed an increase of 120 percent in mass environmental incidents over the previous year.[26] Since the first anti-PX protest in Xiamen in 2007, urban residents in numerous locations across the country have opposed "locally unwanted land uses" (LULUs), including waste incinerators, uranium processing facilities, and chemical plants. The most widely known include the series of demonstrations beginning in 2007 protesting against the PX plants in Fujian Xiamen, in 2011 against such plants in Liaoning Dalian and Yunnan Kunming, in 2014 in Guangdong Maoming, and in 2016 in Shanghai Jin Shan.[27] Protests in cities such as Xiamen have left a considerable digital footprint that can be used by protestors in other cities.

These protests sought to pressure authorities to respond to environmental and social-justice concerns. Rather than merely being examples of "not in my backyard" (NIMBY) campaigns, however, environmental protests in China are frequently predicated upon concerns about limited transparency, public participation, and regulatory incompetence in relation to project siting processes and project operation.[28] Also contributing to the rise of China's environmental activism, public uprisings in China therefore layer together environmental concerns, debates about urban development, responses to unsustainable economic development, and then a raft of questions about the competence and fairness of local governance.

In "Examining Failed Protests on Wild Public Networks: The Case of Dalian's Anti-PX Protests," Elizabeth Brunner examined such a protest, which occurred in Dalian in 2011. Brunner uses the concept of the "wild public screen" to examine how media and people intertwine, how information moves and flows, and how emotional appeals engaged people in and after the anti-PX protest. I appreciate how, as a US-based scholar, Brunner's work is based on extensive fieldwork in China, and how she develops a theoretical concept tailored to China's social movements, which, she argues, are significantly different from a US-style model of "democracy or rational debate." The concept is especially useful to bring our attention back to the fundamentally emotional appeals of China's mass protests and other social movements, which sometimes lack a clear rational component.[29] It would be interesting to compare how environmental-related NIMBY movements develop and conclude in China and the United States differently and to understand their different nature, style, and social and political consequences.

Sometimes, protest can wear a different outfit. In his chapter "Urban Planning as Protest and Public Engagement: Reimagining Mong Kok as an Eco-City," Andrew Gilmore, who has also lived and worked in China, analyzes Vicky Chan's architectural design to see what it suggests about possible modes of protest and public engagement. Focusing on space as a form of environmental communication, Gilmore argues that Chan's design of a Green Mong Kok—his imagined future eco-city, one built in part upon the environmental and democratic spirit of the Umbrella Revolution—creates a site of truncated urban design that fails to take into consideration socioeconomic justice. Gilmore's interesting chapter shows how questions of urban design, the fundamental issue of how public space is organized to facilitate or hamper shared action, impact both our general sense of belonging in a place and the specific possibilities of working together for an imagined environmental future. This is an area of communication scholarship ripe for more attention, as US cities too have been trying to reimagine themselves as "green" and "sustainable" and "eco-friendly." Indeed, the future of US-China relations could be enhanced dramatically if our architects, urban planners, and communication teams would share their expertise on these questions, perhaps envisioning a new global aesthetic of and for green cities.

In addition to protests responding to crisis and these questions of urban development, we saw in the first section of this book how green public cultures have formed around ecotourism, which responds to and cultivates a sense of care. More often than "wild" forms of public participation in specific neighborhoods and cities, Chinese green public culture adopts a leisure-oriented form of peaceful participation through ecotours and trendy "wilderness excursions." Chapters by Li and by Xu offer examples of how environmental communication research might study these practices and perhaps advance a sense of being in and with nature. In "Selling the 'Wild' in China: Ancient Values, Consumer Desires, and the *Quyeba* Advertising Campaign," Xinghua Li traces ancient Chinese discourse on "wilderness excursion" up to the modern influence of Western outdoor culture and industry, providing a case study of how the sublime concept of the "wilderness" travels from ancient times, combines with Western influences, and reshapes the Chinese public's popular hobby of rediscovering nature through outdoor activities. Li shows how conceptions of the "wilderness" appear in various religions in ancient China, and how they have morphed in contemporary China, in part under the influence of Western marketing discourse. While celebrating ecotourism and eco-adventure as ways of submersing the individual in nature, Li is also wary of how such actions may trample nature in the interests of consumerism, making these forms of recreation particularly complicated in terms of what they mean for our green communication.

From a cross-cultural angle, Janice Hua Xu, in "From 'Charmed' to 'Concerned': Analyzing Environmental Orientations of Wildlife Tourists through Chinese and English TripAdvisor Reviews," compared Chinese and English reviews of Mount Emei Natural Ecology Monkey Reserve on TripAdvisor.com. She finds that while the Chinese reviews reflect an anthropocentric view of nature, the English reviews speak of an ecocentric view. Her analysis illustrates how human-nature relations are viewed differently, and how such differences may affect their actions as tourists differently. More of such cross-cultural studies comparing different views and actions on care for nature are needed.

Implications for Further Study

The four tracks I have reviewed herein are by no means comprehensive, but hopefully they point to some fruitful directions of environmental communication that can be examined and explored in future scholarship. Indeed, the voices and materials included in the case studies that make up this book deserve more comprehensive empirical and theoretical investigation; we hope our chapters thus encourage our colleagues to extend this work in new and exciting directions. At the same time, future communication scholarship will want to include other stakeholders—many who are not addressed herein, including Chinese and American business organizations and scientists, international ENGOs, multiparty and transnational environmental organizations, and local advocacy groups in both the United States and China. The field of possible sites of inquiry and engagement is wide open.

We need to continue identifying the many gaps and overlaps in and between US and Chinese environmental communication in terms of the concepts, participants, style of action, their relationships, and so on. For instance, the rich and complex meaning and essence of "nature" and "environment" in the Chinese and US context need to be further explored and compared. Authors from our book such as Li, Xu, Lv and Goodnight, and Kääpä have already explored "environment" in China as cultural, religious, political, and diplomatic resources. For example, Xinghua Li's chapter implies the importance of further examining the concept of "nature," including how its cultural and religious connotations differ in China and the United States, and how such differences might affect our contemporary actions with nature. From a different angle, both Lv and Goodnight's and Gilmore's chapters depict how "environment" serves as a diplomatic tool and political resource, which reshapes the world order in transition and constructs ties between ecologies, countries, and politics as China rises. How are these multiple dimensions and levels of "environment" unpacked differently in the United States? What implications do such differences have upon our actions from initiating green advocacy campaigns to enacting environmental policy both locally and internationally?

Particularly given the theme of the book series in which this book appears, US–China Relations in the Age of Globalization, it is important to stress one gap: the roles of local, regional, and national governments are dramatically different in China and the United States as they impact environmental communication and enact environmental policy. Future scholarship will profit from engaging in cross-cultural analyses of these governing differences, perhaps pointing to best practices wherein each side can learn from the other. It is important to note, however, that even as we strive to engage in the cross-cultural study of environmental communication in and between the United States and China, doing so does not guarantee agreement, consensus, or the development of a unified, global model of environmental action. The realities of international geopolitics, local political machinations, and national heritages will always impinge upon our environmental work.

Nonetheless, I argue that producing more cross-cultural comparative studies will help us to better understand environmental communication and practices in the United States, China, and beyond. For example, imagine a study comparing political debates about the Hetch Hetchy river valley in California to similar controversy over China's Nu River. Imagine a study comparing communication practices around smog in Los Angeles or Salt Lake City or Denver to such discourses depicting Beijing's notorious "airpocalypse" enveloped by fine particulate matter (PM2.5). Imagine a study addressing China's infamous cancer villages to the notorious Superfund sites in America, which likewise have their cancer-causing legacies, silences, and controversies.[30] Across these comparisons, we could hope to learn about the specific and local dimensions of environmental communication, but also about the general and global dimensions, perhaps enabling us to begin to understand how China and the United States share elements of crisis, care, and futurity.

Across these studies, scholars and activists in China and America will continue to debate the fundamental question of how our respective political systems enable, constrain, and compromise our environmental efforts. Particularly in the Chinese context, the very notion of *green public culture* requires further analysis. In previous work that Goodnight and I published (2008, 2016), we proposed looking at China's approach to environmental communication through the lens of green public culture. Empirically, green public culture in China is activated by a mixed green movement of Chinese style, including a government that has been aggressively growing its legislative framework and policy involvement in environmental protection; state organizations and state-sponsored government-organized NGOs (GONGOs) pursuing pro-environmental (and often pro-Party) actions; ENGO advocacy and struggle for surviving and thriving; traditional and new media and individuals speaking up on environmental issues; a Chinese public that is increasingly involved in participation in environmental events, sometimes through mass protests; and continued involvement of international NGOs (INGOs) in extending China's green networks into the global sphere. These diverse players often create multiple discourses and interactions at macro, mesa, and micro levels and form complex and dynamic relationships built on nested networks and activities. Together, these players form an emergent, performative, and complex array of green public cultural spaces composed of environmental discourses, ideas, and power struggles. As is true in the United States and beyond as well, these Chinese green public cultures are shot through with cooperation, contestation, confrontation, collaboration, imagination, and conflict; meanwhile, they are generative, emerging, and constitutive, drawing upon distinct cultural heritages and sociopolitical conditions.[31]

Behind the dazzling green peacock image on our book cover is a quietly brewing controversy over whether the habitat of the endangered green peacock should give way to a dam to be built on the Red River in Yunnan province in China (as discussed in Jingfang Liu and Jian Lu's chapter). At the time of this writing, both advocates for protecting the green peacock and the dam construction company are eagerly waiting for an order to be made from the court, which has been delayed for more than six months. Obviously, it is a hugely difficult decision to be made, while the judge faces tremendous pressure from both sides and a seemingly irreconcilable dilemma amid the *paradoxical modernity*, a term developed by Donovan Conley to describe China's "unprecedented progress borne out of unprecedented environmental calamity."[32] Echoing a similar controversy over whether the snail darter or the Tellico Dam should remain on the Tennessee River, which occurred in the United States more than forty-five years ago, the green peacock debate and its solution are highly symbolic: It speaks of a hard fight between economic development and environmental protection—should the life of a wild species be deemed more valuable than economic gain, or vice versa? This is not just a dilemma that the judge faces, but one that would impact the future of China and beyond equally.[33]

China's environmental crises are creating huge challenges to the living environment, human health, social stability, Party legitimacy, and economics in China. What's more, China's environmental problems are so interconnected with those of other parts of the world—such as Japan, California, Canada, Kenya, and Thailand—with an even more profound transboundary impact globally, that Judith Shapiro comments, "It is within China that much of the future of the planet will be decided."[34] This book presents some of the stories about who are working on these crises, how, and why we care. I hope, however, that the academic research covered by this book is only the first page for both scholars and the public to learn about and explore the much richer landscape of environmental communication in China, the United States, and beyond. Green public cultures in China and beyond emphasize imagination, reflection, and the mutual learning that makes collaboration possible—goals this book hopes to encourage. Indeed, because the environment transcends the borders of nations, so should our understanding: we therefore need to consider international and cross-cultural practices, stories, models, and solutions if we hope to build a sense of environmental care that matches the ever-growing scope of the crises we face. Only by engaging in this mutual learning and sharing, both practically and academically, as advocates and scholars, can we hope to thrive as cosmopolitan citizens. In this fundamental sense, then, we have argued in this book that the environment lies not only at the heart of any thinking we might do about the future of US-China relations, but at the heart of what it means to think about survival in our increasingly sophisticated and networked Anthropocene.

NOTES

1. Thomas Johnson, "Environmentalism and NIMBYism in China: Promoting a Rules-Based Approach to Public Participation," *Environmental Politics* 19, no. 3 (2010): 430–448.

2. John Chung-en Liu and Anthony A. Leiserowitz, "From Red to Green?," *Environment: Science and Policy for Sustainable Development* 51, no. 4 (2009): 32–45.

3. These strategies include China's national goal, set in 2007, to make renewable energy in the country including hydroelectric, biomass, wind, and solar power account for 10 percent of national total energy supply by 2010, and 15 percent by 2020 (see Liu and Leiserowitz, "From Red to Green?," 32–45); the national five-year plan's (2007–2012) ambitious environmental goals to reduce GDP per capita energy intensity by 20 percent (see ibid.); a national strategy, "China's National Climate Change Programme," released by the government in 2007 to address climate change; a shift toward a new green governance approach and a new development model announced in China's current 12th Five-Year Plan (2011–2015), which emphasizes institutional changes to combat with implementation failures in environmental management (see "Tasks and Goals for the 12th Five-Year Plan," *China Daily*, March 6, 2011, http://www.chinadaily.com.cn/china/2011npc/2011-03/06/content_12122766.htm.

4. The construction of ecological civilization has become an important part of building socialism with Chinese characteristics. In 2007, the 17th CPC National Congress first proposed "building ecological civilization" and listed it as one of the new requirements for attaining the goal of building a moderately prosperous society (*xiaokang shehui*) in all respects (see "Full Text of Hu Jintao's Report at 17th Party Congress," *People's Daily Online*, last modified July 5, 2009, http://en.people.cn/90001/90776/90785/6290141.html). In 2012, the 18th CPC National Congress incorporated ecological progress in the five-pronged overall plan for a socialist cause with Chinese characteristics, requiring the nation to make the construction of ecological civilization a foundation for the other four developments on economic, political, cultural, and social fronts in a timely and urgent manner (see "Full Text of Hu Jintao's Report at 18th Party Congress," *People's Daily Online*, last modified November 22, 2012, http://en.people.cn/90785/8024777.html). In March 2018 the Amendment to the Constitution of the People's Republic of China was adopted,

and the "ecological civilization" was officially written into the Constitution. ("Zhonghua renmin gongheguo xianfa xiuzhengan" [Amendments to the Constitution of the People's Republic of China], *Xinhua Net*, last modified January 20, 2019, http://www.xinhuanet. com/politics/2018lh/2018-03/11/c_1122521235.htm).

On August 15, 2005, Xi Jinping, then-secretary of the Communist Party of China Zhejiang Provincial Committee, visited Yucun village in Anji, Huzhou city, and proposed the "Two Mountains" theory, stating that "Lucid waters and lush mountains are invaluable assets." (See "Zhejiang Huzhou: jianshou chulai de meili" (Zhejiang Huzhou: Stick to the beauty), *CCTV.com*, last modified January 20, 2019, http://tv.cctv. com/2018/04/19/VIDEOm9QxhaDu8Tu7LRyl0A6180419.shtml). Nine days later, Xi explained in detail the theory in the article "Clear Waters and Lush Mountains Are Also Invaluable Assets," published by *Zhejiang Daily* (*Zhijiang Xinyu*), and emphasized the importance of building ecological agriculture, ecological industry, ecotourism, and so on, to actualize the theory (see "Lvshui qingshan jiushi Jinshan yinshan" shi da shihua ["Lucid Waters and Lush Mountains are Golden and Silver Mountains" is the truth], *Zhejiang News*, last modified January 20, 2019, http://zjnews.zjol.com.cn/ system/2015/05/31/020676611.shtml.

In 2015 the "Two Mountains" theory was officially written into central documents ("Zhonggong Zhongyang guowuyuan guanyu jiakuai tuijin shengtai wenming jianshe de yijian" [Opinions of the State Council of the Central Committee of the Communist Party of China on Accelerating the Construction of Ecological Civilization], *Xinhua Net*, last modified January 20, 2019, http://www.xinhuanet.com//politics/2015-05/05/c_1115187518.htm).

At the 19th CPC National Congress in 2017, Xi Jinping further emphasized the importance of building ecological civilization in sustaining China's development, and that the two mountains are invaluable assets. Both are key to building the "Beautiful China" initiative and for building global ecological security (see "Full Text of Resolution on Amendment to CPC Constitution," *Xinhua Net*, last modified January 20, 2019, http:// www.xinhuanet.com/english/2017-10/24/c_136702726.htm).

5. "Lianheguo fabu *Zhongguo shengtai wenming zhanlue yu xingdong* baogao" (The United Nations issues the strategy and action report on China's eco-civilization), *People's Daily Online*, July 4, 2016, http://world.people.com.cn/n1/2016/0527/c1002-28383245.html.

6. Shuwen Li, "Huanjing chuanbo de shenshi yu zhanwang" (The Overview and Prospect of Environmental Communication), *Modern Communication* (2010): 39–42.

7. Kalina Oroschakoff and Paola Tamma, "5 Takeaways from the COP24 Global Climate Summit," *Politico*, December 16, 2018.

8. Jianwei Wang and Qichao Wang, "China's Soft Power in Mekong: A Documentary Perspective. Case Study by 'Nourished by the Same River,'" *Polish Journal of Political Science* (2015): 24–52.

9. Dipanjan Roy Chaudhury, "Pushing Back against China's One Belt One Road, India, Japan Build Strategic 'Great Wall,'" *Economic Times*, May 17, 2017. Also see a YouTube debate titled "The Belt & Road Initiative Is a Trillion-Dollar Blunder || Debate #2," https://www.youtube.com/watch?v=fgXmUubTIYw&t=508s. In the debate, some experts argue that the OBOR is for China to play political leverage around the world, that it is more strategic and political than economic, and that it is an ideological competition with the United States. They also point out that small countries are concerned with the risk of signing up for it. Others argue that the OBOR is a historical continuation of the "Silk Road" that will allow Asian countries to knit themselves back together through a massive infrastructure connectivity.

10. Li, "The Overview and Prospect," 39–42.

11. Guanghui Jia, "Zhongguo huanjing xinwen chuabo sanshi nian: huigu yu zhanwang" (30 Years of China Environmental News Communication: Overview and Prospect), *Academic Journal of Zhongzhou*, no. 6 (2014): 168–172.

12. Li, "The Overview and Prospect," 39–42.

13. Xiandu Tao and Youqiong Liu, "Dazhong chuanmei yu nongcun huanjing baohu yulun kongjian jiangou" (Mass Media and the Construction of the Public Opinion Forum for Rural Environmental Protection), *Journal of Hebei Normal University of Science & Technology (Social Sciences)*, no. 9 (2010): 1–5; Jia, "30 Years of China Environmental News Communication," 168–172.

14. For a critical analysis of these urgent issues in current China, see Donovan S. Conley, "China's Fraught Food System: Imagining 'Ecological Civilization' in the Face of Paradoxical Modernity, in *Imagining China: Rhetorics of Nationalism in an Age of Globalization*, ed. Stephen J. Hartnett, Lisa B. Keränen, and Donovan Conley (East Lansing: Michigan State University Press, 2017): 175–205; Lisa Keränen, Kirsten Lindholm, and Jared Woolly, "Imagining the People's Risk: Projecting National Strength in China's English-Language News about Avian Influenza," in Hartnett, Keränen, and Conley, eds., *Imagining China*, 271–303.

15. Jia, "30 Years of China Environmental News Communication," 168–172.

16. Jingrong Tong, *Investigative Journalism, Environmental Problems and Modernization in China* (London: Springer, 2015).

17. Li, "The Overview and Prospect," 39–42.

18. Tong, *Investigative Journalism*, 2015.

19. Xiaoping Guo, "Lun shengtai jilupian kuawenhua chuanbo de youshi, celve jiqi fengxian lunli" (The Strengths, Strategies and Ethical Risks of Cross-Cultural Communication in Eco-documentaries), *The 8th China Visual Art Summit* (2014): 114–120.

20. Songnan Fan, "Cong 'fanxiang yundong' shijiao tanxi huanjing chuanbo zhong de shehui liliang" (The Social Force of Environmental Communication from 'Counter Movement' Perspective), *Modern Communication*, no. 3 (2013): 104–105; Jilong Wang, "Wo guo huan bao fei zheng fu zu zhi de yu lun jian du gong neng yan jiu" (The Public Opinion Surveillance Function of China Environmental NGOs), *Journal of Southwest University for Nationalities* (Humanities and Social Sciences Edition), no. 6 (2013): 173–178.

21. All-China Environment Federation, "Zhongguo Huanbao Minjian Zuzhi Fazhan Bao" (China Environmental Protection NGO Development Report), May 6, 2008, http://www.acef.com.cn/news/lhhdt/2009/0526/9394.html; "Woguo yiyou jin 8000ge huanbao minjian zuzhi" (There Are Nearly 8,000 Environmental Non-governmental Organizations in China), *China Environmental Protection Network*, December 9, 2013, http://www.chinanews.com/ny/2013/12-09/5596400.shtml.

22. Shechuang Xing, "2018 nian Zhongguo cizhanhui shehui qiye renzheng mingdan gongshi" (Public Notification of Social Enterprise Certification List on the China Charity Exhibition in 2018), September 20, 2018, http://www.chinadevelopmentbrief.org.cn/news-21962.html; SheChuang Xing, "2017 nian Zhongguo cizhanhui shehui qiye renzheng mingdan gongshi" (Public Notification of Social Enterprise Certification List on the China Charity Exhibition 2017), December 15, 2017, http://www.chinadevelopmentbrief.org.cn/news-20637.html.

23. Jonathan Sullivan and Lei Xie, "Environmental Activism, Social Networks and the Internet," *China Quarterly*, no. 198 (2009): 422–432; Guobin Yang, "Environmental NGOs and Institutional Dynamics in China," *China Quarterly*, no. 181 (2005): 46–66; Jingfang Liu, "Picturing a Green Virtual Public Space for Social Change: A Study of Internet Activism and Web-based Environmental Collective Actions in China," *Chinese Journal of Communication* 4, no. 2 (2011): 137–166.

24. "Katuoweizi qihou dahui kaimu zhongguojiao jiangshu duoguo minjian zuzhi canyu qihou bianhua gushi" (COP24 in Katowice Opens and China Corner Tells Stories of Multinational NGOs' Participation in Climate Change), *China Climate Act Network*, December 8, 2018, https://mp.weixin.qq.com/s/XfGImjcGTLnw_3io80ExDg.

25. Lei Xie and Hein-Anton Van Der Heijden, "Environmental Movements and Political Opportunities: The Case of China," *Social Movement Studies* 9, no. 1 (2010): 51–68; Jia, "30 Years of China Environmental News Communication," 168–172.

26. Shu Wang, "Jinnian Lai Zhongguo Huanjing Qunti Shijian Gaofa Nianjun Dizeng 29%"

(Environmental Mass Incidents in China Have Increased by 29% Annually in Recent Years), *Beijing News*, October 27, 2012.

27. Yingping Li, "Lun chengshi shequ de huanjing chuan bo" (On the Environmental Communication of Urban Community), *Modern Communication* 36, no. 5 (2014): 38–42.

28. Thomas Johnson, "Good Governance for Environmental Protection in China: Instrumentation, Strategic Interactions and Unintended Consequences," *Journal of Contemporary Asia* 44, no. 2 (2014): 241–258.

29. Xiao'an Guo, "The Choice and Mobilization Effect between Sanity and Emotion in Social Conflicts—Based on Statistical Analysis of 120 Events in 10 Years," *Chinese Journal of Journalism and Communication*, no. 39 (2017): 107–125; Wei Sun, "'Who Are We': The Collective Identity Construct on New Social Movements by Mass Media—Case Analysis on Mass Media Reports of Xiamen PX Project Event," *Journalism Quarterly*, no. 3 (2007): 140–148; Guobin Yang, "Of Sympathy and Play: Emotional Mobilization in Online Collective Action," *Chinese Journal of Communication and Society*, no. 9 (2009): 39–66.

30. PM2.5 refers to tiny particulates in the air that are two-and-one-half microns in width; it is invisible and extremely dangerous to human health. On December 30, 2011, the Ministry of Environmental Protection in China adopted new environmental air-quality standards, including the average PM2.5 concentration and ozone concentration index. On January 21, 2012, the Environmental Monitoring Center of Beijing Environmental Protection Bureau began to publicize its PM2.5 research data on an hourly basis. (See Wenyu Zhu, "PM2.5 shijian yu zhonggguo lvse gonggong lingyu de xingcheng" (PM2.5 Event and the Formation of a Green Public Sphere in China), *Science Popularization*, no. 3 (2015): 40–49; Jianrong Qie, "Huanbaobu: gebie diqu yin huanjing wuran chuxian aizhengcun" (Ministry of Environmental Protection: Cancer Villages Emerge in Certain Areas Due to Environmental Pollution), *Legal Daily*, February 21, 2013, http://gs.legaldaily.com.cn/content/2013-02/21/content_4215724.htm?node=34710.

31. Jingfang Liu and G. Thomas Goodnight, "China's Green Public Culture: Network Pragmatics and the Environment," *International Journal of Communication*, no. 10 (2016): 5535–5557.

32. Donovan S. Conley, "China's Fraught Food System," 177.

33. Kenneth M. Murchison, *The Snail Darter Case: TVA versus the Endangered Species Act* (Lawrence: University Press of Kansas, 2007).

34. Judith Shapiro, *China's Environmental Challenges* (Cambridge: Polity, 2016), 2.

Environmental Communication between Conflict and Performance

Guobin Yang

This outstanding volume maps out an agenda for the study of environmental communication in China within the global context. To my knowledge, this is the first systematic effort of its kind. Phaedra C. Pezzullo's introduction to the volume lays the groundwork for this research agenda by providing a succinct analysis of the thriving field of environmental communication, its social and intellectual origins in the United States, its central concerns with the ethics of crisis and care, as well as Chinese environmental communication practices and theories. She calls for rethinking environmental communication from multinational and transnational perspectives and emphasizes that theories and practices of Chinese environmental communication warrant finding common ground globally and locally.

Pezzullo's analysis of two aspects of Chinese environmental communication, namely, state policy and green publics, is an exemplar of critical hospitality, which offers a cross-cultural perspective that critiques barriers to environmental protection and applauds environmental leadership. Her discussion of the concept of green public spheres associated with my own work is extremely generous and constructive. I am especially indebted to her for pointing out that the focus on greenspeak in our original conceptualization of green public spheres is confined too narrowly to the Habermasian notion of language. Pezzullo cautions insightfully that "we also must remain attuned to embodied, emotional, and contextual communication systems of expression." I would like to use this insight as a jumping board to explore a few implications for environmental communication in China and for communication research more broadly.

If we were to find theoretical resources to complement Habermasian views of rational-critical discourse, and pay heed to "embodied, emotional, and contextual communication systems of expression," we might do well to turn to Bakhtin's theories of the novel and his dialogic theories of language and discourse. For Bakhtin, discourse is "language in its concrete living totality," not an abstraction. The concrete life of the word is necessarily dialogic. It must have an author, and for that reason, it is also necessarily embodied. "Language lives only in the dialogic interaction of those who make use of it." A word is always addressed to another word to another word, creating a dialogue that is unfinalizable.[1]

Bakhtin's dialogic view of language and discourse as "concrete living totality" means that public discourse is fundamentally dramatic because of its living vibrancy. As Ken Hirschkop puts it, in Bakhtin's theories of discourse, "The theatre, rather than the debating chamber, . . . serves as the model for the public sphere: to become public means to be put on stage rather than to assume the podium."[2] In this dramatic model of the public sphere, discourse is embodied, people are actors in interaction with feelings and emotions, and their interactions are dramatic performances. Like all drama, these interactions and performances involve conflict.

For the field of Chinese environmental communication, this dramatic model of the public sphere has two key implications for future research. The first implication is that we might do well to put conflict closer to the center of our analysis. Although my discussion will be limited to Chinese environmental communication, I venture to say that the field of communication and perhaps social-science research more broadly might also do well to put conflict closer to the center. This is not to say that academia has neglected the study of conflict, but rather to point out that the ways of studying conflict in academia sometimes tend to depersonalize and therefore depoliticize conflict.

In making this claim, I am reminded of E. P. Thompson's critique of Raymond Williams, which I believe is all the more relevant today. In 1961, the same year Williams published *The Long Revolution*, E. P. Thompson published a long review of the book, carried in two issues in *New Left Review*. Thompson acknowledged the achievement of the book and its positive reception, and then switched on his critical mode. He said he had real difficulty with Raymond Williams's "tone." What he meant was that both in *Culture and Society* and in *The Long Revolution*, Williams implicitly presented a view of "genuine communication" as genteel communication, or communication "in the language of the academy." "I cannot see that the communication of anger, indignation, or even malice, is any less genuine," Thompson complains. He finds in Williams's works "a procession of disembodied voices," where real social forces are made abstract: "There are no good or bad men in Mr. Williams' history, only dominant and subordinate 'structures of feeling.'" If for Williams, culture means "a whole way of life," for Thompson, it is "a whole way of conflict."[3]

The history of China's environmental communication is one of conflict, in repressed or overt forms. We see this already in the campaigns about animal protection and forestry protection in the mid and late 1990s as well as in the anti-dam campaigns in the early 2000s. We see conflicts even more clearly in the rural pollution-related protests in 2004 and 2005. And as political scientists H. Christoph Steinhardt and Fengshi Wu argue, conflicts in the form of environmental protests have become more proactive and involved broader constituencies since the mid-2000s.[4] Thus, one might argue that environmental issues in China involve "a whole way of conflict."

The field of Chinese environmental politics has in fact consistently focused on various forms of environmental conflicts.[5] In her review essay, Wu highlights three distinct political themes that have concerned many scholars of environmental politics. Under the first theme of the state and environmental governance, we see complex interactions, negotiations, and conflicts due to the fragmented nature of Chinese authoritarianism, tensions between environmental bureaucracies and other regulating bodies, and gaps between policymaking and policy implementa-tion. The second theme of social activism and bottom-up environmental politics reveals conflicts and negotiations between citizens and various levels of state bureaucracies as well as polluting industries. The third theme of environmental diplomacy and international relations points to interactions between Chinese state and non-state actors and international entities.

Anthropologists have produced insightful analyses of environmental conflicts in China, especially in rural areas.[6] For example, in his ethnographic study of the township of Futian in Sichuan province, Bryan Tilt challenges the postmaterialist thesis about the correlation of environmental concerns with growing wealth.[7] Tilt argues that poor villagers in Futian are just as concerned about their environment as the scholars who survey them, but that their concerns are tied to their own social, political, economic, and cultural contexts and expressed in their own ways, which may often elude the understanding of outside observers. Thus, although factory workers bear the brunt of pollution, they downplay the risks to their health and continue their jobs. Tilt calls this behavior "strategic risk repression" and links it to their status as migrants. He explains their risk-taking behavior as a reflection of the cultural meaning of being a migrant—that migrants are seen as risk takers and they can endure harsh conditions for the sake of their families.

All this is not to say that communication scholars have not studied conflicts when they study Chinese environmental communication. Rather, it is to emphasize that the field of Chinese environmental communication might benefit from closer dialogues with other academic disciplines that have already produced a large amount of impressive scholarship on a great variety of environmental issues and forms of conflict.

Of course, as this volume attests, communication scholars have made, and will continue to make, their unique contributions. And here I turn to the second key implication of a dramatic model of the public sphere for Chinese environmental communication research. A dramatic model of the public sphere implies that discourses, events, and actors are parts of a performance, meaning that they are modeled on scripts and conducted to live up to the expectations of an audience. Moreover, performances do not simply represent or reflect reality; they produce new realities—their effects are performative.

In an insightful op-ed in *The Guardian* entitled "China's Green Pledges Are as Deep as a Coat of Paint," Isabel Hilton—puzzling over why, according to a seemingly bizarre piece of news, a bare mountain slope in Southwest China had been painted green at a cost of £30,000—wrote the following: "The episode resonates with a long Chinese tradition of confusing appearance with reality. When, in the 1950s, Mao Zedong decreed that China would overtake Britain in steelmaking within 15 years and that agricultural yields would double with the application of the correct theory, officials all over China reported that the miracle had taken place."[8]

Inventing reality according to a theory, or some other script, is a kind of performance. This kind of performative behavior is a central feature of Chinese politics from the Mao era to the present. Red Guards in the Cultural Revolution, for example, performed an imagined revolution according to the scripts of revolution they had learned.[9] Other examples of performative behavior may include the pervasiveness of political slogans and signs in public spaces past and present. In China today, the promotion of environmental policies and the state agenda of ecological civilization rely heavily on highly visible media campaigns. One of the most influential state-sponsored environmental campaigns in the 1990s happens to be a media campaign called "China Environment March of the Century." Launched in 1993 with the theme of "Declare War on Environmental Pollution," the campaign dispatched reporters from major news agencies to various parts of the country to cover issues related to environment, energy, and sustainable development, thus generating numerous media stories. In its most extreme form, as the *Guardian* story suggests, performative environmentalism of state actors can become literally acts of greenwashing.[10] In recent years, Chinese local government officials in several cities have been caught beautifying their environment and urban landscape by painting barren mountain slopes green, green-spraying yellowing grass along city roads, and installing fake sheep on grassland to make it look as if the grassland were still grazeable.

But it is not just state actors who are engaged in what we might call, following sociologist Jessica Smartt Gullion, performative environmentalism.[11] Citizens and civil society groups also perform, albeit in different ways and for different purposes. China's many environmental NGOs (ENGOs) have staged dramatic and emotional scenes, in the streets or on social media, to protest pollution or polluting industries. Through community projects such as demonstration "greenhouses" built for recycling, they perform civic citizenship to raise environmental awareness and negotiate good relations with government agencies.

I do not imply that performative environmentalism is all for show, or that it is all style and appearance with no substance. On the contrary, I argue that by paying close attention to the performative features of Chinese environmentalism—its styles, scripts, repertoires, rhetoric, and its apparatus of visual presentation—we will gain deep new insights into how an enduring mode of governance and public participation is reinventing itself in the area of environmental communication. How and why does the environment in China become a public arena for dramatic performances, and with what social and political consequences? How much of it has to do with the saturation of media and the Internet in everyday life? Do state and non-state actors communicate about environmental issues differently? What is gained or lost in their respective performances? How much of this performative environmentalism reflects the character of Chinese politics, and to what extent is it a feature of all politics?

Performativity is not peculiar to Chinese culture or society but is part and parcel of the human experience. As sociologist N. K. Denzin puts it, "Performances and their representations reside in the center of lived experience. We cannot study experience directly. We study it through and in its performative representations."[12] In today's media-saturated culture, however, performance merits special attention. Communication scholars who specialize in the study of media, discourse, narrative, rhetoric, stylistics, visual culture, and more, are uniquely equipped to advance this research agenda.

NOTES

1. Mikhail Bakhtin, *Problems of Dostoevsky's Poetics*, trans. Caryl Emerson (Minneapolis: University of Minnesota Press, 1984), 181.

2. Ken Hirschkop, "Justice and Drama: On Bakhtin as a Complement to Habermas," *Sociological Review* 52, no. 1 (2004): 54.

3. E. P. Thompson, "The Long Revolution," *New Left Review* 1, no. 9 (May–June 1961): 24–33.

4. H. C. Steinhardt and F. Wu, "In the Name of the Public: Environmental Protest and the Changing Landscape of Popular Contention in China," *China Journal* 75 (2015): 61–82.

5. Fengshi Wu, "Environmental Politics in China: An Issue Area in Review," *Journal of Chinese Political Science* 14, no. 4 (2009): 383–406.

6. Bryan Tilt, *The Struggle for Sustainability in Rural China: Environmental Values and Civil Society* (New York: Columbia University Press, 2009); Michael J. Hathaway, *Environmental Winds: Making the Global in Southwest China* (Berkeley: University of California Press, 2013); Anna Lora-Wainwright, *Resigned Activism: Living with Pollution in Rural China* (Cambridge, MA: MIT Press, 2017).

7. Tilt, *Struggle for Sustainability.*

8. I. Hilton, "China's Green Pledges Are as Deep as a Coat of Paint," *The Guardian*, February 20, 2007.

9. Guobin Yang, *The Red Guard Generation and Political Activism in China* (New York: Columbia University Press, 2016).

10. Hilton, "China's Green Pledges."

11. Jessica Smartt Gullion, *Fracking the Neighborhood: Reluctant Activists and Natural Gas Drilling* (Cambridge, MA: MIT Press, 2015).

12. N. K. Denzin, *Performance Ethnography: Critical Pedagogy and the Politics of Culture* (Thousand Oaks, CA: Sage Publications, 2003), 12.

About the Authors

Elizabeth Brunner is an assistant professor of communication, media, and persuasion at Idaho State University. US-born, she has studied at both Tsinghua University in Beijing and Nankai University in Tianjin. Brunner has published essays in the *Journal of Communication*, the *International Journal of Communication*, *Critical Studies in Media Communication*, and *Argumentation and Advocacy*, and is author of the book *Environmental Activism, Social Media, and Protest in China: Becoming Activists over Wild Public Networks*, which is based on her ethnographic fieldwork in China.

Andrew Gilmore is a doctoral candidate in the Department of Communication Studies at Colorado State University. His research interests lie in rhetorical theory and criticism, with a particular focus on Hong Kong and its complex relationship with mainland China. Included in his research is the examination of issues surrounding national identity, censorship, cultural preservation, protest, and questions of democracy.

G. Thomas Goodnight is a professor at the Annenberg School for Communication and Journalism at the University of Southern California. He was a Fulbright Senior

Scholar and a visiting scholar at Fudan University and the University of Pennsylvania. During his years at Northwestern University and USC, he has directed over fifty dissertations. Goodnight co-initiated the Earth Science Climate Initiative at USC and hosts a research group on All Things China.

Pietari Kääpä, originally from Finland, is an associate professor in media and communications at the University of Warwick in England. He has published widely on Nordic cinema (including *Ecology and Contemporary Nordic Cinema*, 2014; *Nordic Genre Film*, 2015) and on environmental media politics (including *Environmental Management of the Media: Policy, Industry, Practice*, 2018; *Transnational Ecocinemas*, 2013). He is leading an Arts and Humanities Research Council (AHRC) network titled the Global Green Media Production Network.

Xinghua Li, originally from Shanghai, China, is an associate professor at Babson College, in Massachusetts. She authored *Environmental Advertising in China and the USA: The Desire to Go Green* (2016). Li also has published in journals such as *Media, Culture and Society*, *Communication and Critical/Cultural Studies*, and *Environmental Communication*, and for three years she authored the "Looking Abroad" column for *21st Century*, an English-language weekly of the *China Daily* News Group.

Jingfang Liu, originally from Beijing, is an associate professor at the School of Journalism, Fudan University in Shanghai. Supported by grants from institutions such as the US National Science Foundation and the Chinese Ministry of Education, her research covers environmental communication, organization for strategic communication, information and communication technologies and social change. She has published in leading journals, such as the *Journal of Communication and Critical/Cultural Studies* and the *International Journal of Communication,* as well as chapters in books, such as the *Oxford Handbook of Corporate Reputation.*

Jian Lu, originally from Jiangsu, is an assistant professor at the School of Political Science and Public Administration, China University of Political Science and Law in Beijing. His PhD dissertation investigates the mechanisms driving the emergence of environmental protests in urban China since the 2000s. His research includes contentious politics, environmental governance, and environmental nongovernmental organizations in contemporary China. He has published in peer-reviewed journals, including *China: An International Journal* and *China Journal.*

Junyi Lv is a PhD student at the Annenberg School for Communication and Journalism, University of Southern California, and a visiting scholar at Shanghai Jiao Tong University. Born in Yangquan, Shanxi, China, she has presented her work at major academic conferences internationally and has published in the *International Journal of Communication* and the *International Journal of Cultural Studies*.

Phaedra C. Pezzullo is an associate professor at University of Colorado Boulder. She authored the award-winning *Toxic Tourism* (2007), coedited *Environmental Justice and Environmentalism* (2007), and edited *Cultural Studies and the Environment, Revisited* (2010). In addition to serving on six journal editorial boards, she coauthors the textbook *Environmental Communication and the Public Sphere* (2015, 2018) and coedits the University of California Press book series *Environmental Communication, Power, and Culture*. She lectures throughout the United States and internationally, as well as consults with governments and NGOs.

Binbin Wang is the first PhD on climate change communication in China (Renmin University of China) and the program director at the Institute of Climate Change and Sustainable Development, Tsinghua University, in Beijing. In 2010, Wang cofounded the China Center for Climate Change Communication (China4C), which raises awareness of climate change and promotes innovative strategies. Wang works across media, NGOs, academia, and think tanks, including serving as the special advisor on South-South climate cooperation for the UN.

Janice Hua Xu is an associate professor of communication at Holy Family University in Pennsylvania. She has published widely, including in *Media, Culture & Society*, *Global Media Journal*, and the *International Journal of Communication*. She is a recipient of the Pennsylvania State University Arthur W. Page Center Legacy Scholar Grant for ethics in public communication and served as an expert witness in the US District Court Southern District of New York.

Guobin Yang is the Grace Lee Boggs Professor at the Annenberg School for Communication and the Department of Sociology at the University of Pennsylvania. He authored *The Red Guard Generation and Political Activism in China* (2016), *The Power of the Internet in China: Citizen Activism Online* (2009), and *Dragon-Carving and the Literary Mind* (2003). Yang also edited or coedited four books, including *China's Contested Internet* (2015).

Qinnan (Sharon) Zhou is a researcher at the China Center for Climate Change Communication (China4C) at Peking University and a climate and energy policy freelance consultant. Her previous experiences include project manager at the World Wildlife Fund China in Beijing, strategic partnership assistant for Asia at the Climate Reality Project, and research assistant for the Woodrow Wilson International Center for Scholars in Washington, DC.

Index

Page numbers in italics refer to figures.